Tobias Voigt

Quantitative MR Imaging of the Electric Properties and Local SAR based on Improved RF Transmit Field Mapping

Vol. 11
Karlsruhe Transactions on Biomedical Engineering

Editor:
Karlsruhe Institute of Technology
Institute of Biomedical Engineering

Eine Übersicht über alle bisher in dieser Schriftenreihe erschienene Bände finden Sie am Ende des Buchs.

Quantitative MR Imaging of the Electric Properties and Local SAR based on Improved RF Transmit Field Mapping

by
Tobias Voigt

KIT Scientific Publishing

Dissertation, Karlsruher Institut für Technologie
Fakultät für Elektrotechnik und Informationstechnik, 2010

Impressum

Karlsruher Institut für Technologie (KIT)
KIT Scientific Publishing
Straße am Forum 2
D-76131 Karlsruhe
www.ksp.kit.edu

KIT – Universität des Landes Baden-Württemberg und nationales
Forschungszentrum in der Helmholtz-Gemeinschaft

KIT Scientific Publishing 2011
Print on Demand

ISSN: 1864-5933
ISBN: 978-3-86644-598-7

Quantitative MR Imaging of the Electric Properties and Local SAR based on Improved RF Transmit Field Mapping

Zur Erlangung des akademischen Grades eines

DOKTOR-INGENIEURS

von der Fakultät für

Elektrotechnik und Informationstechnik

des Karlsruher Instituts für Technologie

genehmigte

DISSERTATION

von

Dipl. Phys. Tobias Voigt

geboren in Hamburg

Tag der mündlichen Prüfung: 2. November 2010
Hauptreferent: Prof. Dr. rer. nat. Olaf Dössel
Korreferent: Prof. Dr. rer. nat. Volker Rasche

Abstract

In recent years, quantitative imaging methods gained increasing interest in magnetic resonance imaging. In contrast to standard anatomical imaging, quantitative imaging is expected to improve reliability, accuracy, and reproducibility. This work presents three new methods and their validation and comparison to reference measurements.

Firstly, a new method for simultaneous mapping of B_p and T_1 is developed via a generalization of the standard Actual Flip-angle Imaging (AFI), based on a dual-TR steady-state sequence. An analytical description of the magnetization of a steady-state sequence with an arbitrary number of TRs is derived and forms the basis of the new Multiple TR B_p / T_1 Mapping (MTM). The MTM method outperforms standard AFI in terms of signal-to-noise ratio and B_p mapping accuracy. Both, MTM B_p and T_1 mapping are validated using gold standard references and are successfully applied in *in vivo* experiments.

Secondly, Electric Properties Tomography (EPT), a method for quantitative imaging of dielectric properties of tissue, is presented. The method is based on the measurement of the active component of the excitation field. It is validated in simulation and phantom studies. Based on EPT, a new approach is developed, which allows separate (phase-based) conductivity and (amplitude-based) permittivity measurements. These methods are validated in phantom studies and successfully applied in *in vivo* experiments. Quantitative maps of the human brain are obtained within clinically reasonable imaging times. Particularly phase-based conductivity imaging seems to be mature for clinical evaluation.

Thirdly, a method for patient-specific local specific absorption rate (SAR) measurements is presented. It is based on excitation field measurements and the subsequently reconstructed dielectric properties. Again, the new method is tested and validated in simulation and phantom studies and applied to *in vivo* experiments. Here, local SAR distribution has been measured in the human head, showing good agreement with corresponding, patient-individual simulations.

Preface

Nil actum reputans si quid superesset agendum.

— CARL FRIEDRICH GAUSS (1777 - 1855)

This thesis describes the research I carried out as part of my PhD study at the Karlsruhe Institute of Technology in Karlsruhe, Germany, and at the Philips Research Laboratories in Hamburg, Germany. During my studies and working on the topics of this thesis, I received a tremendous amount of support by many people. I am deeply indebted to all these people and without their help, I would have never been able to complete this thesis.

First and foremost, I would like to thank my supervisor and professor Olaf Dössel from Karlsruhe Institute of Technology and my supervisor Ulrich Katscher from Philips Research Laboratories. Olaf supported me in every aspect, from questions regarding the scientific work and discussion on the topic of dielectric properties to all sorts of organizational issues like planned publications, proposals for additional funding, or coordinating the defense of my dissertation. Special thanks go to Ulrich for teaching me how MRI works and for endless discussions on literally all aspects of MRI. Ulrich initiated this project. I could learn and benefit a lot from his vast experience in MRI research and I especially enjoyed our brainstorming sessions about new ideas and projects.

I also like to thank Dye Jensen and Gerhard Krohn for giving me this unique opportunity to join the Tomographic Imaging group of Philips Research in Hamburg and to carry out my PhD thesis there.

The largest part of the work described in this thesis was carried out at Philips Research Laboratories. With its numerous experts on nearly all aspects of MRI, it is an ideal environment to undertake research and address all kinds of scientific questions regarding medical imaging. My work could benefit a lot from this innovative research environment and I am very thankful to everyone who helped me during these three years.

I would like to devote special thanks to the members of the EPT project;

to Christian Stehning for teaching me many practical and important tweaks regarding the scanner operation, Christian Findeklee for providing and explaining the simulation tools and coil models and for thoroughly revising all the analytical calculations, and Kay Nehrke for his valuable support on field mapping. I would like to express my gratitude also to all other people involved in MR research, including Stefanie Remmele, Julien Sénégas, Jochen Keupp, Ingmar Graesslin, Oliver Lips, Peter Vernickel, Holger Eggers, and Peter Börnert for triggering new ideas, asking critical questions, and for solving all sorts of problems. From the Digital Imaging group I received support from Ewald Rössl who introduced me to Cramér-Rao lower bound calculations and from Martin Bergtholdt who helped me with image segmentation.

Let me also say 'thank you' to my fellow PhD candidates and already finished PhDs who shared the "John Malkovich floor". I enjoyed matches of table football with Ronny and Uwe, coffee breaks with Mariya, discussions about football with Alfonso, and regular running around the Alster with Eberhard, Sebastian, and Hanno.

Some calculations being part of this thesis have been conducted in Karlsruhe. Although only being an occasional guest in Karlsruhe, I always felt very welcome and a part of the research group. I would like to especially thank Sebastian Seitz and Julia Bohnert for providing the FDTD field data I used for testing nearly every new idea before performing actual experiments.

During my work I could benefit from a scientific collaboration with the University Medical Center in Utrecht. I would like to thank Astrid van Lier and Nico van den Berg for inspiring and motivating discussions.

Last but not least I would like to thank Ingelore and Burkhard for their support throughout all these years. Finally, I thank Kerstin for always being there.

Tobias Voigt
Hamburg, September 2010

Contents

CHAPTER ONE

INTRODUCTION

When you can measure what you are speaking about, and express it in numbers, you know something about it; but when you cannot measure it, when you cannot express it in numbers, your knowledge is of a meagre and unsatisfactory kind: it may be the beginning of knowledge, but you have scarcely, in your thoughts, advanced to the stage of science, whatever the matter may be.

— LORD KELVIN (1824 - 1907)

1.1 General introduction

Magnetic resonance imaging (MRI) is an imaging technique used primarily in medical settings to produce high quality, soft tissue contrast images of the inside of the human body. The importance of tomographic imaging for medical applications such as diagnostics as well as therapy assessment has gained increased interest during the last decades. Non-invasive imaging modalities such as MRI with its high innovative power and outstanding level of detail are also often used as a tool in medical research. The main areas of application of MRI lie in the fields of neurologic imaging and musculoskeletal imaging. Other applications are given by breast and body imaging as well as combining MRI with interventional procedures.

In the last years, quantitative imaging techniques based on MRI have gained increasing popularity. Usually, MR image voxels are characterized by gray values with arbitrary units. The relative difference in gray values is perceived as image contrast and used for medical applications. The MR scanner is used as a "camera" that produces pictures. In quantitative imaging, not only the relative image contrast, but also its absolute scale bears information. Quantitative imaging represents a shift of paradigm towards using the MR scanner as a scientific measurement instrument.

In standard anatomical imaging, macroscopic structures are compared with previous examinations or healthy individuals. The quality of information gained from the new quantitative way of MRI can be strikingly different. Quantitative parameters are able to reflect specific structural changes of human tissue and cells (see Section 1.4 of this introduction). Thus, using quantitative imaging techniques, direct access to changes of the underlying biochemical processes of the human organism is enabled and can be used to investigate changes caused by disease well below anatomical MR resolution.

Ideally, quantitative imaging techniques are independent of many factors influencing the technical process of MR image acquistion, and thus, provide better reproducibility and comparability. Consequently, quantitative parameters should be independent of the actual scanner used, the field strength, small changes in the sequence protocol, and - most importantly - independent of the radiologist's individual interpretation. Typical quantitative imaging techniques used in clinical environments are perfusion and diffusion imaging as well as relaxometry and the subsequent determination of, e.g., contrast agent concentration, vessel size, or partial oxygen pressure. These techniques will be described briefly in the introductory chapter of this work. Further chapters will introduce additional, not yet explored quantitative imaging parameters and / or techniques. A corresponding literature review will be given in the introductory parts of each chapter.

In Chapter 2, a new technique for simultaneous mapping of the amplitude of the excitation field and the longitudinal relaxation time constant is presented. The method is inspired by an existing method to map the magnitude of the radio frequency (RF) transmit field, however, providing a more general theoretical basis and overcoming its shortcomings. The new method yields a higher signal-to-noise ratio (SNR) per unit time and is also free of systematic errors from relaxation effects, encountered frequently in excitation field mapping techniques. Vice versa, its relaxation time mapping aspect is independent of excitation field inhomogeneities as often encountered in stand-alone techniques mapping relaxation times.

Chapter 3 describes MR measurements of two new quantitative parameters, the electric conductivity and permittivity, i.e., the dielectric properties of the patient's tissue. The underlying basic theory is elaborated from basic physical principles. In MRI, the dielectric properties of human tissue distort the excitation field. Based on a mapping of these distortions as described in the previous chapter, Maxwell's electro-dynamic equations are arranged to yield a reconstruction formula for dielectric properties. The new imaging technique is tested in simulations and phantom experiments. Finally, *in vivo* feasibility is demonstrated and, for the first time ever, quantitative conductivity and permittivity values of the human brain are obtained *in vivo* with MRI.

Chapter 4 is based on the methods presented in the last two chapters. Maps of the excitation field and the dielectric tissue properties are used to calculate the local specific absorption rate (SAR). The local SAR is a position dependent measure of the rate at which energy is absorbed by the body when exposed to an RF field. RF induced tissue heating is a major problem for MRI at high field strengths ($\geq 3\,\mathrm{T}$), particularly for the upcoming parallel RF transmission systems. Highly demanding imaging sequences could lead to too high local energy deposition, and eventually, to tissue burning. To prevent this situation, regulatory committees in the EU and USA defined upper limits of global and local SAR, which have to be validated prior to each scan execution. Unfortunately, up to now, no suitable methods to determine global and local SAR *in vivo* have been developed. Instead, simulations based on standardized patient models are applied, and large safety margins are required to take into account differences between the modelled and the individual, actual patient. In this work, a new approach for local SAR estimation is developed, tested in simulations, and applied to *in vivo* experiments. It allows assessing the local SAR in a fast and patient individual manner.

The presented work aims to deliver a proof of concept and a feasibility study of these new MRI methods. Unlike those presented in the introductory sections 1.4.1 - 1.4.5, they do not reflect routine MRI in today's clinical practice. However, a considerable achievement is expected from any application of the presented techniques in a clinical setting, which should be investigated in dedicated trials.

1.2 Magnetic resonance imaging - brief historical perspective

The physical principle underlying MRI is the resonant excitation of spins in a magnetic field. Isidor Rabi was the first to discover the magnetic resonance phenomenon in molecular beams in 1938 [1]. Felix Bloch and

Edward Purcell independently expanded his work to liquids and solids in 1946 [2, 3]. Nuclear magnetic resonance is a property of magnetic nuclei, which are situated in a magnetic field and irradiated by an electromagnetic (EM) RF pulse. The nuclei absorb energy from the RF pulse and radiate this energy back out. Rabi received the Nobel price in 1938 and Bloch and Purcell were both awarded the Nobel Prize for this discovery in 1952. Nuclear magnetic resonance is the basis for magnetic resonance imaging. In the period between 1950 and 1970, NMR was, however, exclusively developed and used for chemical and physical molecular analysis.

In 1971, Raymond Damadian showed that the time when the RF energy is radiated back out from the nuclei, i.e., the nuclear relaxation times of healthy and malignant tissues are different. This motivated scientists to consider magnetic resonance for the detection of disease [4], albeit not yet capable of producing images. In 1973, the X-ray-based computerized tomography (CT) was introduced by Hounsfield [5]. In contrast to the work conducted by Damadian, CT was able to produce images of the inside of the human body. Allan MacLeod Cormack independently invented a similar process, and both Hounsfield and Cormack shared the Nobel price for their discoveries in 1979.

Magnetic resonance imaging was first demonstrated on small test tube samples filled with water by Paul Lauterbur in 1973 [6]. Image reconstruction was first based on a back projection technique similar to that used in CT. In 1975, Richard Ernst proposed MRI using phase and frequency encoding by applying magnetic field gradients in connection with image reconstruction via Fourier Transformation [7]. This technique is the basis of current MRI as used routinely in hospitals. A few years later, in 1977, Raymond Damadian demonstrated MRI of a live mouse called field-focusing nuclear magnetic resonance [8]. In this same year, Peter Mansfield developed the (EPI) technique [9]. This technique will be developed in later years to produce MR images at video rates. The first human MRI was produced in 1977 [10]. In 2003, the Nobel Prize in Physiology or Medicine was awarded to Paul Lauterbur and Sir Peter Mansfield for their discoveries concerning MRI. The physical principles behind MRI and some of its applications are described in the next sections.

1.3 Basic introduction to magnetic resonance imaging

1.3.1 Physical principles

The human body consists primarily of fat and water containing many hydrogen atoms, leading to an abundance of approximately 60% hydrogen atoms throughout the human body. Inherent to each hydrogen nucleus, in fact each elementary particle, is a property called spin. The first direct experimental evidence of the spin of an electron was found in the Stern-Gerlach experiment conducted in 1922 in Frankfurt. Stern and Gerlach found that silver atoms containing one unpaired electron are deflected when exposed to a magnetic field gradient as if they contain a quantized magnetic moment with only two possible orientations. In a mathematical theory of physics, spins can be introduced in the relativistic quantum theory of Paul Dirac. In this framework, particles with spin $\frac{1}{2}$ are described by so called spinors. Particles with spin $\frac{1}{2}$, e.g. electrons, can, when measured, only be encountered in either spin up or down states and hence Stern and Gerlach observed two distinct spots of silver atoms. Apart from this quantization, spins obey the same mathematical laws as do classical angular momenta, and hence for better imagination are often described as the rotation of a particle around itself. Another fundamental difference to classical rotation is the magnetic moment associated with every spin

$$\hat{\mu} = \gamma\hat{\mathbf{J}} = \widetilde{\gamma}\left(g_S\hat{\mathbf{S}} + \hat{\mathbf{L}}\right) \tag{1.1}$$

where $\hat{\mathbf{L}}$ is the angular momentum operator, $\hat{\mathbf{S}}$ is the spin operator, $\widetilde{\gamma} = \frac{\mu_{\mathrm{B}}}{\hbar}$ the gyromagnetic ratio with the Bohr magneton μ_B, and g_S is the g-factor. A circumflex ˆ indicates a quantum mechanical operator. For electrons, $g_s = 2$ follows from Dirac's theory and yields the magnetic moment associated to spin. It is not unity as could be expected from rotation of a classical object. In the following, $\gamma = \frac{g_S\mu_N}{\hbar}$ will be defined for arbitrary nuclei. For neglegible angular momentum, total magnetic moment is caused from spin $\hat{\mu} = \gamma\hat{\mathbf{S}}$ with $\hat{S}_i = \frac{\hbar}{2}\hat{\sigma}_i$ and $\hat{\sigma}_i$ being the Pauli matrices

$$\hat{\sigma}_x = \begin{pmatrix} 0 & 1 \\ 1 & 0 \end{pmatrix}, \quad \hat{\sigma}_y = \begin{pmatrix} 0 & -i \\ i & 0 \end{pmatrix}, \quad \hat{\sigma}_z = \begin{pmatrix} 1 & 0 \\ 0 & -1 \end{pmatrix}.$$

The time evolution of quantum mechanical systems is described by the Schrödinger equation $\hat{H}\psi = E\psi$, where the wave function ψ describes the quantum mechanical state and $\hat{H}$ is the Hamiltonian operator which "extracts" the energy E from the wave function. The Hamiltonian usually consists of a kinetic energy part and a potential energy term. When placed in a magnetic field of strength $\mathbf{B}$, energy states with different spin will experience a Zeeman effect leading to additional Hamiltonian $\hat{H}_Z = -\hat{\mu}\mathbf{B}$. In the following the magnetic field will be assumed to be one-dimensional, pointing in z-direction $\mathbf{B} = B_0\mathbf{e}_z$ and hence $\hat{H}_Z = -\gamma\hat{S}_z B_0 = -\gamma B_0 \frac{\hbar}{2}\begin{pmatrix}1 & 0\\ 0 & -1\end{pmatrix}$. The time dependent Schrödinger equation becomes

$$i\hbar\frac{\partial}{\partial t}\begin{bmatrix}\Psi_+\\ \Psi_-\end{bmatrix} = -\gamma\frac{\hbar}{2}\begin{bmatrix}B_0 & 0\\ 0 & -B_0\end{bmatrix}\begin{bmatrix}\Psi_+\\ \Psi_-\end{bmatrix} \tag{1.2}$$

where $\Psi_{+,-}$ represent spin up and down states. From solving the Schrödinger equation, the energy eigenvalues are obtained $\langle\Psi|\hat{H}|\Psi\rangle = E_{m_s} = -m_s\hbar\gamma B_0$ with $m_s = \pm\frac{1}{2}$ whether the spin is oriented up or down. Hence, in such a magnetic field, an otherwise degenerate state will split up into $2s+1$ states, where for hydrogen it is $s = \frac{1}{2}$ and hence two energy levels. These energy levels are separated by energy gap $\Delta E = E_{-\frac{1}{2}} - E_{\frac{1}{2}} = \gamma\hbar B_0$, a hydrogen atom inside a magnetic field can therefore absorb a photon of a specific energy $\hbar\omega_0 = \gamma\hbar B_0$. The frequency depends on the gyromagnetic ratio and the magnetic field strength as

$$\omega_0 = \gamma B_0 \tag{1.3}$$

ω_0 is called the resonance frequency or the Larmor frequency. For hydrogen, the gyromagnetic ratio is $\gamma/2\pi = 42.58\,\text{MHz/T}$. At typical main field strength B_0 as used in clinical MRI (0.5 T to 7 T), this corresponds approximately to frequencies of 21 to 300 MHz.

To predict the time evolution of the magnetic moment of a spin exposed to the field $\mathbf{B} = B_0\mathbf{e}_z$, the expectation value of the $\hat{\mu}$-operator has to be computed (see e.g. [11])

$$\begin{aligned}
\langle\Psi|\hat{\mu_x}|\Psi\rangle &= \frac{\gamma\hbar}{2}\sin(\theta)\cos(\varphi_0+\omega_0 t)\\
\langle\Psi|\hat{\mu_y}|\Psi\rangle &= \frac{\gamma\hbar}{2}\sin(\theta)\sin(\varphi_0+\omega_0 t)\\
\langle\Psi|\hat{\mu_z}|\Psi\rangle &= \frac{\gamma\hbar}{2}\cos(\theta). \qquad (1.4)
\end{aligned}$$

This Eq. 1.4 describes a magnetization vector $\mu = (\mu_x, \mu_y, \mu_z)^T$ rotating about the z-direction at Larmor frequency ω_0 with cone-angle θ. Without external perturbations, the z-magnetization μ_z is constant over time. The instantaneous phase of transverse magnetization components is given by $\varphi(t) = \varphi_0 + \omega_0 t$. The magnetic moment of spin is used in MRI to generate tomographic images of the human body.

MRI measures the expectation value of the net magnetic moment $\mathbf{M}$ of a very large number of proton spins $\mathbf{M} = (M_x, M_y, M_z)^T = \frac{1}{V}\sum_i \mu_i$. At temperatures of the human body, thermal energy will lead to a stochastically distribution of spin up and down orientations, as well as phase distributions, leading to a vanishing net magnetic moment. At high magnetic fields however, the Zeeman term to the Hamiltonian will increase compared with the thermal energy of the spins and a net magnetic moment, i.e. equilibrium magnetization M_0, can be created. The net magnetic moment will therefore depend on temperature T and main field strength B_0. It can be shown (see, e.g. [12]) that

$$M_0 = \frac{N\gamma^2\hbar^2 B_0}{4kT} \qquad (1.5)$$

where N is the number of spins per unit volume and k is Boltzmann's constant. Hence, an abundance of spins predominantly aligned to one direction will lead to a macroscopic magnetic moment in that direction. In summary, if a patient is exposed to the main magnetic field of an MR scanner, the spins of the hydrogen tend to align parallel to the main magnetic field and begin to precess at Larmor frequency, resulting in a net magnetic moment along the main field direction. These spins can be excited by the incidence of resonant radiation. This process is described in the next section.

1.3.2 Spin excitation

To create a measurable signal emitted by protons, these protons first have to be excited using resonant electromagnetic fields with Larmor frequency ω_0. Excitation in this context means flipping the spin orientation with respect to the main magnetic field. To flip a spin, an additional time dependent magnetic field $\mathbf{B}_1$ will be applied, resulting in overall field $\mathbf{B}(t) = \mathbf{B}_0 + \mathbf{B}_1 = B_0\mathbf{e}_z + B_p(\mathbf{e}_x \cos\omega_0 t + \mathbf{e}_y \sin\omega_0 t)$. The constant over time part $\mathbf{B}_0 = B_0\mathbf{e}_z$ is oriented in z-direction and causes the precession described in the previous section. The time dependent part is responsible for spin flipping. B_p is the amplitude of a positively rotating, resonant RF transmit field $\mathbf{B}_1$.[1] Inserting this into the Schrödinger Eq. 1.2 yields

$$i\hbar\frac{\partial}{\partial t}\begin{bmatrix}\Psi_+\\ \Psi_-\end{bmatrix} = -\gamma\frac{\hbar}{2}\begin{bmatrix}B_0 & B_p e^{i\omega_0 t}\\ B_p e^{i\omega_0 t} & -B_0\end{bmatrix}\begin{bmatrix}\Psi_+\\ \Psi_-\end{bmatrix} \tag{1.6}$$

with ω_0 being the Larmor frequency and $\omega_1 = \gamma B_p$ being the frequency associated with the $\mathbf{B}_1$ field and moving the magnetization vector away from equilibrium. Again, the expectation values of the magnetic moment operator can be computed, yielding [12]

$$\begin{aligned}\langle\mu_{x'}(t)\rangle &= \langle\mu_{x'}(0)\rangle\\ \left\langle\mu_{y'}(t)\right\rangle &= \left\langle\mu_{y'}(0)\right\rangle\cos\omega_1 t + \langle\mu_z(0)\rangle\sin\omega_1 t\\ \langle\mu_z(t)\rangle &= -\left\langle\mu_{y'}(0)\right\rangle\sin\omega_1 t + \langle\mu_z(0)\rangle\cos\omega_1 t\end{aligned} \tag{1.7}$$

where dashed coordinates describe a frame of reference that rotates with frequency ω_0. Eqs. 1.7 describe a vector of fixed magnitude that rotates about the x-axis in the rotating frame of reference. If the $\mathbf{B}_1$ field is turned on for a specific amount of time τ_{RF}, the flip angle α can be controlled by the amplitude B_p via

$$\alpha = \omega_1\tau_{rf} = \gamma B_p\tau_{rf}. \tag{1.8}$$

Here, it has been assumed that the amplitude B_p is constant over time (also called a hard pulse). Hence, for inducing a 90° spin flip $\gamma B_p t = \pi/2$ has to

[1] An idealized description of $\mathbf{B}_1$ is used in this section. In Chapters 3 and 4 a more realistic description is used, which also contains non-vanishing counter rotating B_m and z-components B_z.

be ensured. Scanner hardware is designed to apply the necessary field B_p to achieve the desired spin flip homogeneously over the entire field of view. Additional interactions with the sample will however alter B_p as a function of position $\mathbf{r}$. These alterations can be measured using specific mapping techniques. A method for B_p mapping is described in this introduction in section 1.4.1 and an improved method for B_p mapping is developed in chapter 2. Alterations of $B_p(\mathbf{r})$ encountered in *in vivo* imaging also bear information on the physiology of the sample. This is the basis for a new image contrast presented in chapter 3.

Apart from spin tipping, the energy transferred via the applied high frequency $\mathbf{B}_1$ field also induces heat. The amount of heat that is allowed to be transferred to a human undergoing MRI is regulated by government agencies. In the USA for example, standards are set by the Food and Drug Administration (FDA) allowing a temperature increase of 1° over 6 min investigation time [13]. Limits are expressed as a maximum for local specific absorption rate (local SAR). The deposition of energy during MR scanning is a problem especially at high-field MRI. Local tissue heating can not only occur on the surface of the human body but also inside where not always temperature sensation is possible. Therefore, local SAR has to be assessed before scanning. Chapter 4 provides a new method for patient individual local SAR assessment based on B_p mapping.

The Bloch equation describing the time evaluation of a bulk magnetization vector under the influence of an arbitrary time-dependent field $\mathbf{B}(t) = B_x(t)\,\mathbf{e}_x + B_y(t)\,\mathbf{e}_y + B_z(t)\,\mathbf{e}_z$ can be derived from the time derivative of the expectation values of the magnetization operator [12]

$$\frac{\partial}{\partial t}\begin{bmatrix} M_x(t) \\ M_y(t) \\ M_z(t) \end{bmatrix} = \gamma \begin{bmatrix} 0 & B_z(t) & -B_y(t) \\ -B_z(t) & 0 & B_x(t) \\ B_y(t) & -B_x(t) & 0 \end{bmatrix} \begin{bmatrix} M_x(t) \\ M_y(t) \\ M_z(t) \end{bmatrix}. \tag{1.9}$$

In real spin systems encountered in human tissue, interactions will lead to energy dissipation and eventually to return of magnetization to its thermal equilibrium. In MRI, two constants are used to describe this phenomenon of relaxation.

1.3.3 Relaxation

Excited spins will spontaneously or stimulated emit photons and eventually relax back to the lower energy state. This rate of relaxation will depend on the interplay of the nucleus with its surroundings. It is found that this rate is different for different tissue types. This is the foundation of (standard) MR image contrast. In MRI, relaxation of excited spins is characterized by two time constants. The term T_1 models the decay of M_z and T_2 models the decay of M_t, i.e. the magnetization in te x-y plane.

At equilibrium, the net magnetization vector lies along the direction of the applied magnetic field B_0 and is called the equilibrium magnetization M_0. In this configuration, the z-component of magnetization M_z equals M_0 and is referred to as the longitudinal magnetization. By exposing the spin to electromagnetic fields of resonant frequency ω_0 it is possible to excite the spin and to flip all magnetization into the transverse plane according to the relations described in the last paragraph Eq. 1.7. The spins will however not remain in that position. Interactions between the spin and lattice will lead to a decrease of higher energy level population and hence a return to the equilibrium magnetization. The time constant which describes how M_z returns to its equilibrium value is called the spin lattice relaxation time (T_1). The equation governing this behavior as a function of the time t is the Bloch Equation for longitudinal magnetization

$$M_z(t) = M_z(t_0)\, e^{-\frac{(t-t_0)}{T_1}} + M_0 \left(1 - e^{-\frac{(t-t_0)}{T_1}} \right). \tag{1.10}$$

Examples for MRI-based techniques for T_1 determination are outlined in section 1.3.3 of this introduction. A new method developed in the framework of this thesis is presented in Chapter 2.

After spin flipping about an angle of 90°, the net magnetization vector lies in the transverse plane. The net magnetization vector can be thought of as a superposition of many individual spin vectors. As pointed out in context with Eq. 1.4, spins in a magnetic field precess around the direction of the field starting with some initial phase φ_0 and evolving over time according to $\omega_0 t$. In addition to the rotation, the transverse magnetization starts to dephase because of spin-spin interactions. Each of the spins forming the magnetization vector is experiencing random phase changes due to the interaction with its surrounding spins. The longer the elapsed time, the larger the phase difference between the spins, and hence the lower the detectable signal. The time constant which describes the return to equilibrium of the transverse magnetization (i.e., zero magnetization), is called the spin-spin relaxation time, T_2

$$M_t = M_t(t_0)\, e^{-t/T_2}. \tag{1.11}$$

T_2 is always less than or equal to T_1. On top of the just described spin-spin interaction, a second component accelerating transverse magnetization dephasing are macroscopic B_0 inhomogeneities. Spins experiencing a magnetic field different from B_0 will rotate at frequencies slightly different from ω_0. This leads to an additional decay constant $T_2^{'}$, sometimes called the inhomogeneous effect. The combination of these two factors is what actually results in the decay of transverse magnetization $1/T_2^* = 1/T_2 + 1/T_2^{'}$. In contrast to T_2, $T_2^{'}$ decay can be reversed by, e.g. spin echo sequences (see below).

The Bloch equation Eq. 1.9 can be rewritten taking into account relaxation effects

$$\frac{\partial}{\partial t}\begin{bmatrix} M_x \\ M_y \\ M_z \end{bmatrix} = \begin{bmatrix} -1/T_2 & \gamma B_z(t) & -\gamma B_y(t) \\ -\gamma B_z(t) & -1/T_2 & \gamma B_x(t) \\ \gamma B_y(t) & -\gamma B_x(t) & -1/T_1 \end{bmatrix}\begin{bmatrix} M_x \\ M_y \\ M_z \end{bmatrix} + \begin{bmatrix} 0 \\ 0 \\ M_0/T_1 \end{bmatrix}. \tag{1.12}$$

An application of the Bloch equation to a sequence of RF pulses following rapidly within a pulse repetition time (TR) smaller than T_1 is given in section 2.2.1.

1.3.4 Signal production

In MRI a combination of RF pulses and gradient fields, called a sequence, can be used to generate a signal from relaxing spins. Three main types of signal generation sequences will be introduced in this section. In principle all kinds of pulse and gradient combinations are possible and will lead to different signal and image contrast.

If the magnetization is flipped into the transverse plane and not modified further, the resulting signal is called a free induction decay (FID) and shown in Fig. 1.1a. If the $\mathbf{B}_1$ field is turned off and all magnetization has been flipped into the transverse plane, the protons immediately begin to re-radiate the absorbed energy. As discussed in the previous section, transverse magnetization will decline because of dephasing according to decay constant T_2^* in Eq. 1.11. Longitudinal magnetization will recover according to longitudinal relaxation constant T_1.

A spin echo (SE) sequence [14] is depicted in Fig. 1.1c. A 90° pulse can be used to flip the magnetization into the transverse plane. Dephasing will begin, resulting in a signal decay according to T_2^*. Now, a second pulse is applied inducing a 180° spin flip. The magnetization vector is mirrored with respect to the transverse plane. This pulse will also invert the dephasing direction, eventually leading to a new alignment of all spins and hence leading to a maximum in transverse magnetization. This is called an echo. The time between the applied RF pulse and the center of the echo is the echo time (TE). The amplitude of the echo will decay with time constant T_2, since the dephasing component due to external B_0 field inhomogeneities is refocused. An arbitrary number of 180° pulses can be added to create more echoes [15] (turbo spin echo (TSE)), limited by T_2 decay and by local SAR, which might impose restrictions due to the large amplitude required to induce a spin inversion (cf. Eq. 1.8).

Another way to create an echo is by using gradients (Fig. 1.1b). A gradient after the initial spin flip leads to stronger dephasing due to this additional field inhomogeneity. If the gradient is inverted, the direction of additional dephasing is inverted as well, leading to a signal maximum when all spins are aligned again. This is called a gradient echo. Also in this sequence, multiple echoes can be acquired following one single excitation. For typical combinations of T_1 and TR, higher SNR can be obtained using smaller flip angles than 90°. Typical implementations of gradient echo sequences such as the fast field echo (FFE) can also involve additional gradients and RF pulses to spoil remaining transverse magnetization prior to the next application of the excitation pulse (see also chapter 2 and section 2.4.5).

1.4 Quantitative MRI

MR images may show an almost infinite variety of tissue contrasts, depending on the sequence parameters chosen. It is obvious that conclusions derived from MRI are hence subjective and heavily dependent on radiologist's interpretation. Quantitative information can be used to avoid this problem. Inherent to the paradigm shift towards MR scanning as scientific

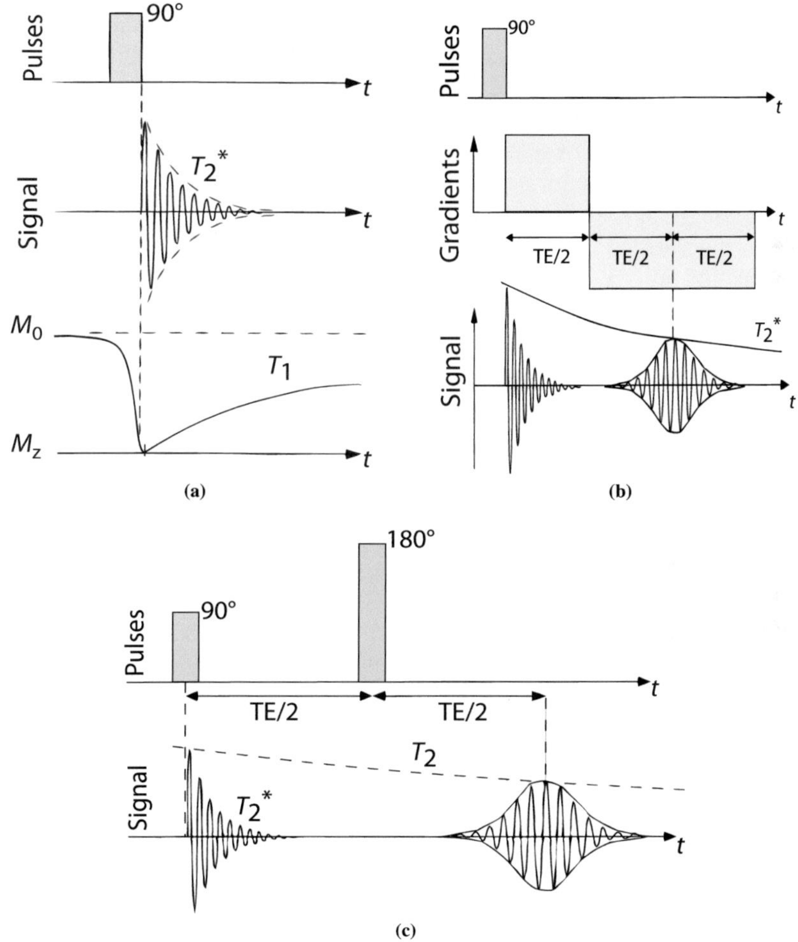

Figure 1.1: (a) Free induction decay. Following the 90° pulse, the transverse magnetization decays according to T_2^*. (b) Gradient echo sequence. An echo is created by gradients used to dephase and realign spins. (c) shows a spin echo sequence. A 180° pulse is used to refocus transverse magnetization leading to an echo. Images are based on [16].

measurements are expectations about strengthened reliability, accuracy, and reproducibility. Quantitative imaging could be very helpful in large multi-center clinical trials [17], since it facilitates standardization of therapy assessment. In this context, it can also be argued that e.g. drug trials could profit from quantitative data analysis allowing conclusions at earlier stages of the trial and hence be less cost-intensive. Another advantage of quantitative imaging over standard anatomical MRI is given by the chance to directly detect structural changes well below anatomical MR resolution.

Although for each quantitative parameter a variety of mapping methods exists, this introduction presents only one method for each parameter for the sake of brevity. This will include a method for mapping of the excitation field, which is also the topic of chapter 2, a method for mapping of the two relaxation times discussed in the previous paragraph T_1 and T_2 of which the former will also be addressed in chapter 2, and probably the most widely used clinical quantitative mapping applications, diffusion and perfusion imaging. For more details on the presented and other methods an overview can be found in the textbook [18].

1.4.1 Mapping of the spin excitation field B_p

RF inhomogeneity is the largest source of error in quantitative MRI ([18], p 22.). Therefore, B_p mapping is frequently used in connection with other quantification techniques, to correct for errors due to spatially varying excitation. The real flip angle is however a quantitative parameter on its own. Usually it is represented as percentage of a constant nominal flip angle, which is given in degree. Sometimes it is also expressed as peak magnetic field strength of the excitation pulse in units of Tesla. In chapter 3, new quantitative parameters are introduced based on mapping of the excitation field, hence changing the use of B_p mapping as a mere correction for other measurements towards providing a link to cell physiology.

Many methods to measure B_p have been proposed. Most of these are briefly discussed in section 2.1. In spite of the large number of proposed methods, the “final”, optimal method does not seem to be found yet, reasoning further developments in this field. Thus, in chapter 2 of this work, the refinement of a recently introduced B_p mapping technique is presented. Another example of recently introduced B_p mapping techniques is the Bloch Siegert B_p mapping [19], which seems to be promising in terms of SNR per imaging time. The Bloch Siegert shift describes the effect of incident non-resonant radiation on precessing spins [20]. Such a pulse will, if shaped properly and its off-resonance is strong enough, lead to a shift of the precession frequency $\omega_0 + \omega_{BS}$ without spin excitation. This off-resonance will then induce a phase shift that depends on the magnitude of the positively rotating component of the RF pulse (B_p) according to

$$\mathrm{d}\varphi_{BS} = \int_0^T \omega_{BS}(t)\,\mathrm{d}t = \int_0^T \frac{(\gamma B_p(t))^2}{2\omega_{RF}(t))}\mathrm{d}t \tag{1.13}$$

where ω_{RF} is the off-resonance frequency and $\mathrm{d}\varphi_{BS}$ is the Bloch Siegert phase shift. Acquiring two phase images with and without Bloch Siegert off resonance pulse, the phase shift can be measured and Eq. 1.13 can be used to determine B_p. Instead of performing two subsequent measurements, two off-resonant pulses within one pulse sequence can be used to eliminate contributions from inhomogeneous B_0. Typical off-resonances used in Bloch Siegert B_p mapping are $4-8\,\mathrm{kHz}$ applied for $8\,\mathrm{ms}$ [19].

1.4.2 Mapping of the longitudinal relaxation time T_1

The relaxation constant of longitudinal magnetization is a quantitative parameter and measured in milliseconds. The gold standard for T_1 measurements is a series of inversion recovery experiments. It consists of an initial $180°$ inversion pulse that flips the entire magnetization in the negative z-direction, followed by an excitation pulse (e.g., $90°$) at different inversion recovery (IR) delays (t_{IR}) .

The inversion pulse prepares the longitudinal magnetization, which subsequently recovers exponentially according to T_1 (cf. Eq. 1.10). The excitation pulse then flips the amount of longitudinal magnetization that has recovered at t_{IR}, causing an FID according to Eq. 1.11. The amplitude of this FID will depend on the longitudinal relaxation time T_1. Using different IR delays, the T_1 recovery curve can be sampled fitting the measured signals to the theoretical expression

$$S(t_{IR}) = S_0\left(1 - 2e^{-t_{IR}/T_1}\right) \tag{1.14}$$

where $\mathrm{TR} \gg T_1$. Long repetition times have to be used to ensure that the magnetization has fully recovered before the next inversion. Technically

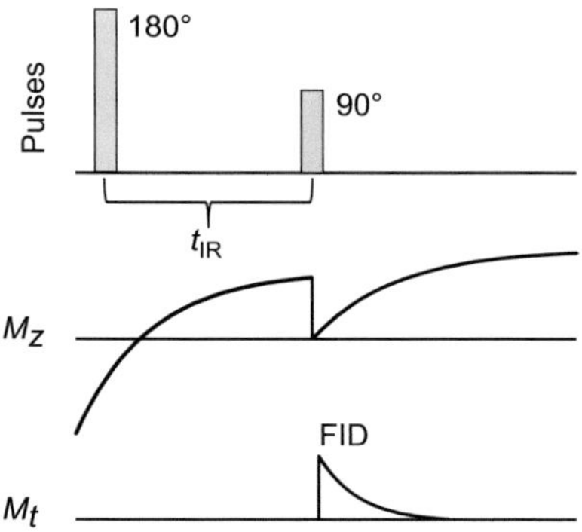

Figure 1.2: Inversion recovery T_1 mapping sequence. Initially the magnetization is inverted and then recovers according to T_1. The 90° pulse causes an FID that can be detected. A single application of the pulse sequence is displayed.

this sequence can be implemented as a single shot technique, i.e. using multiple spin echoes in one FID and acquiring different imaging slices in an interleaved manner, drastically increasing imaging speed despite the required long TR. A further increase of imaging speed can be achieved by applying different t_{IR} in a single sequence [21]. In this work, a new method to measure T_1 (and simultaneously B_p) is presented in chapter 2 and compared to a single shot inversion recovery sequence. At 3 T, typical T_1 relaxation times in the human body are of the order of hundreds of milliseconds up to a few seconds.

1.4.3 Mapping of the transverse relaxation time T_2

Quantitative T_2 maps can be generated by acquiring a series of images with different TEs. As TE is growing, the signal intensities will decrease exponentially (cf. Eq. 1.11 with t replaced by TE_i)

$$S_i(t) = S_0 e^{(-\mathrm{TE}_i/T_2)}. \tag{1.15}$$

If several images are acquired at different time points, a pixel by pixel exponential fit to the measured data can be used to determine T_2. To create these images, a Carr-Purcell-Meiboom-Gill (CPMG) sequence is usually applied [22]. In a CPMG sequence, the spins are tipped into the transverse plane using a 90° pulse and then refocused using a 180° pulse with

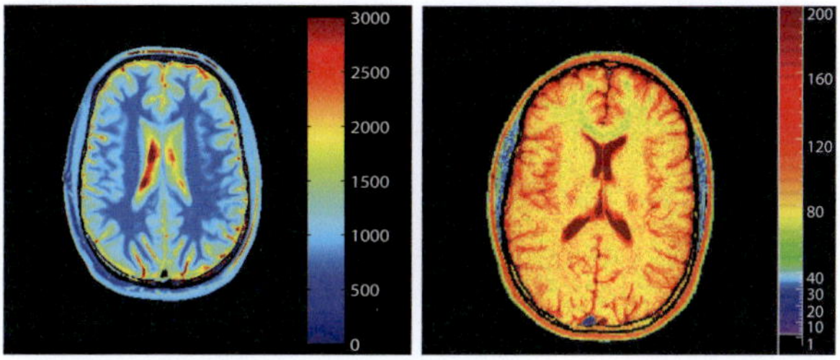

(a) **(b)**

Figure 1.3: (a) T_1 map obtained using a inversion recovery sequence in units of ms. Courtesy Mariya Doneva. (b) T_2 map obtained using a CPMG sequence in units of ms. Courtesy Julien Sénégas.

a 90° phase offset. As pointed out before, the RF excitation field is influenced by a variety of parameters, e.g. the dielectric properties of the sample. Therefore, planned flip angles will not necessarily be reproduced exactly over the image leading to e.g. non-perfect inversion pulses. Via the CPMG technique with alternating phases, errors in the 180° pulses can be corrected and every second (even) echo will represent exact T_2 decay and hence used for fitting. At 3 T, typical T_2 relaxation times in the human body are of the order of $5-200$ ms. A T_2 map of the brain is shown in Fig. 1.3b. When gradient echo sequences are used to sample the decay curve, T_2 has to be replaced by T_2^* in Eq. 1.15.

1.4.4 Diffusion imaging with MR

In diffusion imaging, the molecular mobility of water in tissue is measured. The diffusion depends on tissue microstructure and can therefore provide crucial information on the cellular organization of healthy and diseased tissue. The destruction of biological barriers for example will have immediate influence on the diffusion of water. The diffusion coefficient D describes the relation between macroscopic flux density J_i and concentration gradient $\nabla_i C$, both are directionally dependent

$$J_i = D\nabla_i C \tag{1.16}$$

where i indicates spatial direction. D is given in units of mm^2/s. In MRI diffusion imaging, displacements of magnetized spins are of the order

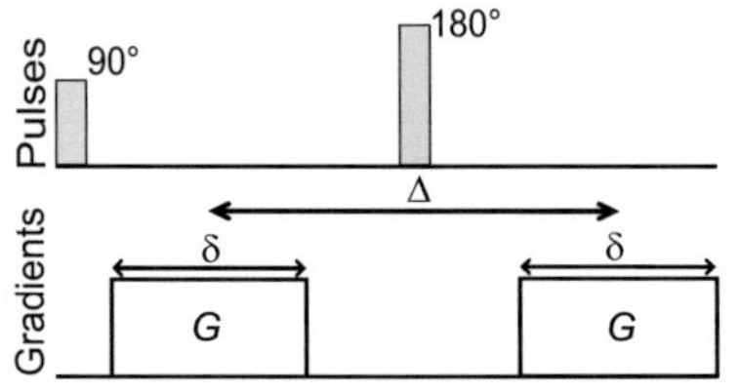

Figure 1.4: Diffusion imaging sequence. A spin echo experiment is combined with two gradient pulses. Spins that have diffused between the two gradient pulses will experience a phase difference.

of millimeters and time periods of the order of minutes. For simplicity, the first derivations employ homogeneous media in which the mobility of spins shows no directional dependence and hence scalar D. D is also called the apparent diffusion coefficient (ADC).

Diffusion information can be encoded into MR images by using two gradient pulses in combination with a spin echo experiment. Such a sequence is depicted in Fig 1.4. The first gradient pulse is applied directly after the 90° spin flip. It induces a position dependent phase shift $\gamma G\delta a$ between two spins that are separated by distance a. δ is the length of the diffusion sensitizing gradient and G its strength. Now, the magnetization is inverted by the 180° pulse and the same gradient is applied again. For static spins this will accurately refocus the effect of the first gradient pulse. For diffused spins, a net phase difference will remain. This will lead to signal attenuation for diffused water. The ratio of a diffusion weighted image to a standard spin echo is given by the Stejskal Tanner equation [23]

$$\ln\left(\frac{S(G,\Delta,\delta)}{S(0)}\right) = -\gamma^2 G^2 \delta^2 \left(\Delta - \frac{\delta}{3}\right) D \tag{1.17}$$

where $S(G,\Delta,\delta)$ and $S(0)$ are the magnitudes of the MR signal with and without applied diffusion gradients. Δ is the time interval between the two gradients. Usually, the effects of gradient strength and timing are summarized into the b-factor with units $\mathrm{s/mm^2}$

$$S(b) = S(0)\exp(-bD). \tag{1.18}$$

The diffusion constant D can now be obtained by acquiring diffusion weighted images with varying b values and fitting the results on a pixel by pixel basis to Eq. 1.18.

The diffusion direction and hence the value of the diffusion coefficient as described here is clearly defined by the direction of the gradients relative to the sample. A reproducible quantification of diffusion that does not depend on e.g. patient position inside the scanner can be found using difffusion tensor imaging (DTI). In general, $\mathbf{D}$ is a symmetric second order tensor

$$\mathbf{D} = \begin{pmatrix} D_{xx} & D_{xy} & D_{xz} \\ D_{xy} & D_{yy} & D_{yz} \\ D_{xz} & D_{yz} & D_{zz} \end{pmatrix}$$

where D_{xy} represents the diffusion along y when a concentration gradient along x is present. The diffusion tensor can be measured using different directions of gradient pulses. The resulting matrix $\mathbf{D}$ can be diagonalized and its eigenvectors represent the principal directions of diffusion. The corresponding eigenvalues will correspond to diffusion values along these directions regardless how the patient was oriented during the scan. For more details on diffusion tensor imaging refer to [18].

1.4.5 Perfusion imaging with MR

The parameters that can be obtained in perfusion imaging characterize the blood flow at the capillary level and are closely related to the delivery of oxygen and other nutrients to the tissue. It has to be distinguished from bulk blood flow of the large blood vessels. Clinical perfusion imaging has been used for e.g. classification of tumors or the identification of acute stroke regions. Bolus tracking and arterial spin labeling (ASL) are two approaches that have been developed to measure perfusion with MR. In the following, bolus tracking is summarized. For a review of ASL techniques, please refer to [23].

For bolus tracking, a bolus of contrast agent is injected into the vascular system and the effect on the tissue relaxivity is measured during the passage of the bolus via rapid imaging. The contrast agent reduces the T_2, T_2^*, and T_1 of the tissue and hence will lead to signal attenuation during passage through the volume of interest (VOI). In this introduction, SE-based perfusion imaging will be described and hence the T_2 lowering of a dedicated T_2 contrast agent will be monitored (tracked). This method is

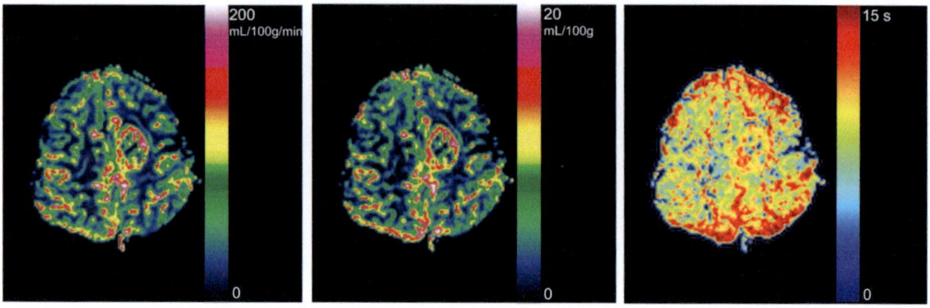

(a) (b) (c)

Figure 1.5: (a) shows an image of CBF, measured in milliliters of blood per 100 gram of tissue per minute. (b) shows an image of CBV measured in milliliters of blood per 100 gram of tissue. (c) shows an image of MTT measured in seconds. Images by courtesy of Prof. Atle Bjørnerud, PhD (The Interventional Centre, Oslo University Hospital & Dept. of Physics, University of Oslo, Oslo, Norway).

called dynamic susceptibility contrast MRI (DSC-MRI)[2]. T_2 induced signal changes are believed to give better indication of perfusion although the signal change in T_2^* weighted sequences is larger. The concentration of tracer C_{VOI} is a function of time and can be linked to the rate of change of signal intensity in MR

$$C_{\mathrm{VOI}}(t) = k_{\mathrm{VOI}} \Delta R_2(t) \tag{1.19}$$

with change in relaxivity $\Delta R_2 = \Delta(1/T_2)$ and k_{VOI} being a tissue and pulse sequence dependent parameter. On the other hand, in a model for non-diffusable tracers and intact blood-brain barrier the concentration of tracer can be linked to physiologic parameters

$$C_{\mathrm{VOI}}(t) = \frac{\rho}{k_{\mathrm{H}}} F_{\mathrm{VOI}}(C_a(t) \otimes R(t)) \tag{1.20}$$

where $\otimes$ denotes a convolution and $R(t)$ is the residue fraction of contrast agent still present in the VOI following an ideal injection, and C_a is the arterial input function (AIF). C_a describes the concentration of contrast agent entering the VOI. ρ is the density of tissue and k_{H} is the haematocrit

[2]In the literature T_2 dynamic contrast enhanced MRI (DCE-MRI) and DSC-MRI are often used interchangeably.

factor. F_{VOI} is the blood flow. The residue $R(t)$ can be related to the transport function $h(t)$ describing the distribution of transit times of the contrast agent, which is related to the vascular structure and flow

$$R(t) = 1 - \int_0^t h(\tau)\,\mathrm{d}\tau. \tag{1.21}$$

The relation of Eqs. 1.19 and 1.20 yields physiological parameters that are accessible using DSC-MRI. Via a measurement of changes in relaxivity ΔR_2 , the following quantities that characterize brain perfusion can be computed:

- Cerebral blood flow (CBF) (F_{VOI}). CBF measures the rate at which blood is delivered to the tissue. By deconvolving Eq. 1.20 the quantity $F_{\text{VOI}} \cdot R(t)$ can be obtained. Using that initially $R(0) = 1$, F_{VOI} can be obtained. Accurate quantification of CBF however requires an accurate description of the AIF (C_a), which is usually not straight forward. The unit of CBF is milliliters of blood per gram of tissue per minute (ml/g/min). Fig. 1.5a shows an example of CBF.

- Cerebral blood volume (CBV). CBV is the volume of blood in the voxel per mass of tissue within the voxel

$$\text{CBV} = \frac{k_H}{\rho} \frac{\int C_{\text{VOI}}\mathrm{d}t}{\int C_a \mathrm{d}t}. \tag{1.22}$$

  CBV is measured in units of ml/g. The normalization by arterial input in Eq. 1.22 can be understood since the more tracer is injected, the more will eventually reach the VOI. Fig. 1.5b shows an example of CBV.

- Mean transit time (MTT). MTT describes the average time interval needed for a particle of the tracer to pass through the vascular bed within the VOI. Fig. 1.5c shows an example of MTT. MTT is measured in seconds (s)

$$\text{MTT} = \frac{\int t h(t)\,\mathrm{d}t}{\int h(t)\,\mathrm{d}t}. \tag{1.23}$$

These three quantities can be connected via the central volume theorem $\text{MTT} = \frac{\text{CBV}}{F_{\text{VOI}}}$. Next to the described parameters, a variety of summary parameters not directly related to physiology are often calculated. These describe features of the measured signal such as time to peak (TTP), bolus arrival time (BAT), maximum peak height (MPH) or full-width at half-maximum of the bolus (FWHM). Typical problems encountered in perfusion quantification using DSC-MRI are the aforementioned characterization of arterial input. Moreover, it has to be made sure that the influence of the contrast agent itself on perfusion and T_1 is negligible.

CHAPTER TWO

SIMULTANEOUS MAPPING OF B_p AND T_1 USING MULTIPLE REPETITION TIME STEADY-STATE GRADIENT ECHO SEQUENCES

Scienza è il distinguere quello che si sa da quello che non si sa.

— GALILEO GALILEI (1564-1642)

Abstract — This chapter presents a new approach towards a fast and simultaneous mapping of the amplitude of the radiofrequency (RF) transmit field (B_p) and the longitudinal relaxation time T_1. The new method is based on the actual flip angle imaging (AFI) sequence. However, the single pulse repetition time (TR) pair used in the standard AFI sequence is replaced by multiple TR sets. The resulting method was coined multiple TR B_p/T_1 mapping (MTM). The chapter starts with the derivation of an analytical expression for the signal of an RF-spoiled steady-state sequence with arbitrary number of repetition times. The properties of MTM are investigated and compared to standard AFI in simulations and experiments. Feasibility and reliability of MTM are proven in phantom and *in vivo* experiments. Error propagation theory is applied to identify optimal sequence parameters and to facilitate a systematic noise comparison to standard AFI. In terms of accuracy and signal-to-noise ratio, the presented method outperforms standard AFI B_p mapping over a wide range of T_1. Finally, the capability of MTM to determine T_1 is analyzed qualitatively and quantitatively, yielding good agreement with reference measurements.

Based upon:

- T. Voigt, K. Nehrke, O. Dössel, U. Katscher; T_1 Corrected B_1 Mapping Using Multi-TR Gradient Echo Sequences. *Magnetic Resonance in Medicine* 2010;64:725-733.
- T. Voigt, S. Remmele, U. Katscher, O. Doessel; Fast T1/B1 Mapping Using Multiple dual-TR RF-spoiled Gradient Echo Sequences. In *Proceedings of the 18th Annual Meeting of ISMRM*, Stockholm, Sweden 2010. p. 2949.

2.1 Introduction

In MRI applications, spatially varying sensitivity profiles of the RF transmit coils and local variations of the conductive and dielectric properties in the tissue may result in a spatially inhomogeneous amplitude of the RF transmit field B_p.[1] Therefore, accurate B_p mapping is required for a variety of applications. For instance, non-uniform B_p distributions are an essential source of errors in quantitative imaging. Thus, B_p mapping is frequently used for post-correction of, e.g., spin density [24], T_2 [25], T_1 [26], or magnetization transfer measurements [27]. Moreover, B_p mapping allows the estimation of the electric conductivity and local SAR (see Chapters 3 and 4). In the framework of parallel transmission, RF shimming [28] and accelerated local excitation via transmit sensitivity encoding (SENSE) [29] rely on accurate B_p field information. Here, the need for sequential mapping of the B_p distribution of multiple transmit channels makes particularly fast B_p mapping techniques highly desirable. Several approaches towards fast and accurate B_p mapping have been developed in the last decades.

The B_p transmit field can be measured via the double angle method [30], which has been refined repeatedly to increase accuracy and/or reduce scan time [31], [32–34]. Furthermore, separate images based on multiple flip angles are used for fitting suitable signal equations by [35, 36]. Another approach comprises the application of stimulated echoes in multi-pulse sequences, where signals with different flip angle dependence are acquired [37–39]. The study presented in [40] suggests using the 180° signal void to map B_p fields. A recently published B_p mapping method uses two RF pulses to encode the flip angle information into the spatial phase distribution [41]. A B_p mapping method based on the Bloch Siegert effect due to off-resonant RF has been introduced in 2010 [19]. Finally, actual flip angle imaging (AFI) maps the B_p transmit field using a fixed flip angle in a dual TR steady-state gradient echo sequence [42]. In recent years, AFI has gained increasing popularity, because it represents a fast and potentially robust B_p mapping approach. Several studies have been reported concerning different aspects of AFI, e.g., spoiling of transverse magnetization components [43,44]. A recent study compares AFI to other

[1] B_p is used to denote the amplitude of the positively rotating, circularly polarized component of the spatial vector of the excitation field $\mathbf{B}_1$. In the literature also the notations B_1 or B_1^+ are used for B_p.

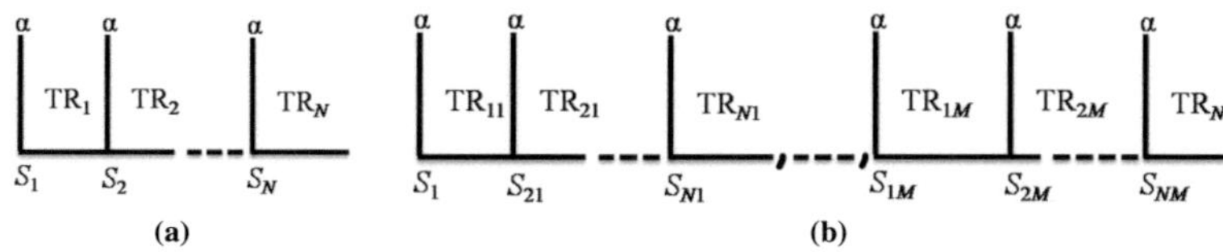

Figure 2.1: Sketch of a steady-state sequence based on N different repetition times TR_n is shown in (a). Sketch of M sequences containing N different repetition times TR_{nm} is shown in (b). Each pulse is applied with the same flip angle α.

B_p mapping methods [45].

Besides speed and SNR, systematic errors caused by T_1 relaxation may be a problem in accurate B_p mapping. Some approaches suggest using repetition times much longer than T_1 to eliminate these errors [30, 34, 36], however, leading to impractical long imaging times. Other approaches are independent of T_1 [31, 32, 39–41]. The AFI approach relies on a linearization of T_1 relaxation, and thus, produces a systematic error in the flip angle estimation that depends on T_1. On the other hand, a variety of quantitative T_1 mapping techniques involves post-correction for B_p field inhomogeneities via additional measurements of the transmit field as shown in [24, 26, 46]. A recently published study provides a method for concurrent B_p and T_1 mapping based on analyzing signals from two pulse trains with different initial magnetization [47].

In this chapter, a generalization of the AFI approach is presented, which is based on multiple TR interleaved in a single steady-state sequence and/or distributed over successive scans [48]. Multiple TR B_p/T_1 mapping (MTM) allows the simultaneous determination of both B_p and T_1 maps from the acquired data. Hence, the resulting B_p maps are corrected automatically for T_1-induced errors and vice versa. Due to the aimed derivation of electric properties from B_p maps, the main focus of this study is placed on the B_p mapping capabilities of MTM, regarding the simultaneously acquired T_1 map as correction for the B_p map. The B_p mapping performance of the new method is compared with standard AFI mapping in the framework of error propagation theory. Furthermore, phantom and *in vivo* measurements supporting the new method's superiority regarding accuracy are presented. The T_1 mapping capability of MTM is compared to gold standard inversion recovery T_1 mapping.

2.2 Theoretical motivation for a new combined B_p/T_1 mapping method

2.2.1 Signal equation

A steady-state RF-spoiled gradient echo sequence based on N different, arbitrary repetition times TR_n, $n \in \{1 \ldots N\}$, is depicted in Fig. 2.1a. Apart from the variable delay at the end of each profile (i.e., the variable TR), all sequence parameters are fixed. The corresponding gradient echo signals S_n can be derived from the normalized, transverse steady-state equilibrium magnetization $M_{t,n}$ directly after the nth RF pulse

$$S_n = \exp\left(\frac{-\mathrm{TE}}{T_2^*}\right) M_0 M_{t,n} \tag{2.1}$$

with the equilibrium magnetization M_0. Subsequently applying the Bloch equation (see also Section 1.3.3) for the z-component of the magnetization $M_z(t) = M_z(t_0)\, e^{-\frac{(t-t_0)}{T_1}} + M_0\left(1 - e^{-\frac{(t-t_0)}{T_1}}\right)$ yields an analytical expression for $M_{t,n}$, i.e. the magnetization reached right after the nth pulse reads

$$M_{t,n} = \sin(\alpha) \frac{\sum_{j=1}^{N}\left[\cos^{(N-j)}(\alpha)\left(1 - E_{\mathrm{mod}(j+n-2,N)+1}\right)\prod_{k=j+1}^{N} E_{\mathrm{mod}(j+n-2,N)+1}\right]}{1 - \cos^N(\alpha)\prod_{j=1}^{N} E_j} \tag{2.2}$$

with $\mathrm{mod}(i,N)$ being the modulus function with basis N and $E_n = e^{(-\mathrm{TR}_n/T_1)}$. The analytical expressions Eqs. 2.1 and 2.2 are based on the assumption that there is no remaining transverse magnetization at the time of the next pulse. RF and gradient spoiling should be used to minimize contributions of spurious spin echoes to the signal [43,44]. In the case of $N = 1$, Eq. 2.2 yields the standard steady-state normalized equilibrium equation [12]

$$M_t = \frac{\sin(\alpha)(1 - E_1)}{(1 - \cos(\alpha) E_1)} \tag{2.3}$$

In the case of $N = 2$, Eq. 2.2 yields the dual TR signal equation as used for AFI [42]

B_p mapping method	Number of TRs N	Number of measurements M	Reconstruction technique
(Standard) AFI	2	1 ($M > 1$ used to denote averaging multiple identical measurements)	analytically via signal ratio
Generalized AFI	≥ 3	1 ($M > 1$ used to denote averaging multiple identical measurements)	analytically via signal ratio
MTM	≥ 2	$\geq 1 (N \cdot M \geq 3)$	numerically via model fitting

Table 2.1: Overview of B_p mapping techniques discussed in this study: ''Standard'' AFI described in [42], Generalized AFI introduced in this study, and MTM, also introduced in this study. The cases $M = 1$, $N = 3$ can be reconstructed alternatively via Generalized AFI or MTM.

$$M_t = \sin(\alpha) \frac{1 - E_{2,1} + \left(1 - E_{1,2}\right) E_{2,1} \cos(\alpha)}{\left(1 - \cos^2(\alpha) E_1 E_2\right)} \tag{2.4}$$

In general, various ways can be considered to extract the desired quantities B_p (assumed to be proportional to α) and T_1 from the signals given in Eqs. 2.1, 2.2. In this chapter, two approaches are discussed:

1. a general analytical method based on signal ratios, and

2. a numerical method fitting measured data to Eqs. 2.1 and 2.2.

An overview of the discussed methods is given in Table 2.1.

2.2.2 General analytical method

This section can be seen as a generalization of the work published in [42]. To determine the actual flip angle α, the M_0 and T_2^* dependence in 2.1 can be removed via the ratio $r_{i,j}$ of two signals S_i and S_j

$$r_{i,j} = \frac{S_i}{S_j} = \frac{M_{z,i}}{M_{z,j}}. \tag{2.5}$$

Combining Eqs. 2.5 and 2.2 yields

$$\sum_{k=1}^{N} \cos^{N-k}(\alpha_{i,j}) \left[r_{i,j} \mathrm{TR}_{mod(k+j-2,N)+1} - \mathrm{TR}_{mod(k+i-2,N)+1} \right] = 0. \quad (2.6)$$

Here, $\mathrm{TR}_n/T_1 \ll 1$ was assumed, which allows approximating the exponential factors E_n in Eq. 2.2 by first order Taylor series. This causes a systematic T_1-dependent error present in the analytical approach, and thus, also in AFI B_p mapping. Equation 2.6 is a polynomial of grade $N-1$ of $\cos(\alpha)$ and does not contain any other unknown quantities. In general, if all TR_n values in the sequence are different, $\binom{N}{2}$ ratios (Eq. 2.5) and polynomials (Eq. 2.6) can be computed. Each polynomial has $N-1$ solutions for $\alpha_{i,j}$, which are not all physically meaningful. The case $N=2$ yields one polynomial with one solution, reproducing the results from [42],

$$\alpha_{2,1} \approx \arccos\left(\frac{Rr_{2,1}-1}{R-r_{2,1}}\right) \quad (2.7)$$

with $R = \mathrm{TR}_2/\mathrm{TR}_1$. In the case $N=3$, the solution of Eq. 2.6 can also be derived analytically. For the ratios $r_{i,j} = S_i/S_j$, the physical solutions are found to be

$$\begin{aligned} \alpha_{i,j} \approx \arccos & \left[-\frac{r_{i,j}\mathrm{TR}_{\mathrm{mod}(j,N)+1} - \mathrm{TR}_{\mathrm{mod}(i,N)+1}}{2\left(r_{i,j}\mathrm{TR}_{\mathrm{mod}(j-1,N)+1} - \mathrm{TR}_{\mathrm{mod}(i-1,N)+1}\right)} \right. \\ & + \left(\frac{\left(r_{i,j}\mathrm{TR}_{\mathrm{mod}(j,N)+1} - \mathrm{TR}_{\mathrm{mod}(i,N)+1}\right)^2}{4\left(r_{i,j}\mathrm{TR}_{\mathrm{mod}(j-1,N)+1} - \mathrm{TR}_{\mathrm{mod}(i-1,N)+1}\right)^2} - \right. \\ & \left. \left. -\frac{r_{i,j}\mathrm{TR}_{\mathrm{mod}(j+1,N)+1} - \mathrm{TR}_{\mathrm{mod}(i+1,N)+1}}{r_{i,j}\mathrm{TR}_{\mathrm{mod}(j-1,N)+1} - \mathrm{TR}_{\mathrm{mod}(i-1,N)+1}} \right)^{\frac{1}{2}} \right]. \end{aligned} \quad (2.8)$$

The three solutions Eq. 2.8 are different with respect to systematic errors induced by the relaxation time T_1 and can be used for averaging, lowering mean statistical errors. For $N > 3$, no analytical solution of Eq. 2.6 can be provided. In the following experiments and simulations, only the case $N = 2$ (standard AFI) and $N = 3$ (called generalized AFI) are considered and compared with the numerical fitting approach outlined in the next section.

2.2.3 MTM fitting

Alternatively to the analytical method outlined above, a numerical fitting approach, dubbed MTM, is presented. Here, the flip angle α can be obtained by fitting the measured data to the theoretically expected signals Eqs. 2.1 and 2.2. Additionally, this fit also yields the corresponding T_1 map. The mathematical structure of the signals reads

$$S(\mathrm{TR}_1 \ldots \mathrm{TR}_n, \mathrm{TE}, T_2^*, T_1, \alpha, M_0) = S'(T_2^*, \mathrm{TE}, M_0)\, M_t(\mathrm{TR}_1 \ldots \mathrm{TR}_n, \alpha, T_1)\,. \tag{2.9}$$

Thus, the fitting algorithm requires at least three images to determine three unknown parameters S', T_1, and α. S' acts as a global scaling and will not be used in MTM. These three data points can be obtained by using at least $N = 3$ different repetition times, or by using a $N = 2$ sequence and repeat the measurement M times with varying TR values (see Fig. 2.1b), yielding $K = N \cdot M$ data points. A corresponding, second index m is introduced with $m \in \{1 \ldots M\}$ for the repetition times TR_{nm} and the signals S_{nm}. Thus, also the $K = 4$ data points obtained from $N = M = 2$ are sufficient to determine the three unknowns S', T_1, and α. Of course, arbitrary N and M with $K \geq 3$ are possible. The fitting is performed comparing the K measured signals S_{nm} to the analytical expressions given in Eqs. 2.1 and 2.2. In the next section, it will be shown that MTM reconstructions, i.e., fitting $K \geq 3$ data points, achieve a substantial increase of the reconstruction accuracy of α. Moreover, the assumption $\mathrm{TR}_n \ll T_1$ is not required in this approach, and thus, the reconstructed α is free of systematic, T_1 induced errors.

In the following, the index $m \in \{1 \ldots M\}$, introduced for MTM, will be used also in connection with standard and generalized AFI imaging with $N = 2$ and $N = 3$, respectively. In this case, identical TR sets will be applied M times, corresponding to averaging generalized AFI results, to facilitate the comparison of the different considered schemes.

2.3 Methods

2.3.1 Implementation

The MTM reconstruction of simulated and measured B_p mapping data was performed using in-house C++ software based on the Levenberg-Marquardt algorithm [49].

2.3.2 Performance analysis

The noise behavior of MTM fitting and the generalized analytical approach for $N = 2$ (standard AFI) and $N = 3$ (generalized AFI) was analyzed by means of simulations. Since the goal of this chapter is a T_1 corrected B_p map, this analysis was restricted to B_p performance only.

The comparison of AFI and MTM comprises a systematic and a statistic error. First, AFI suffers from a systematic error $\Delta\alpha$ due to neglecting T_1 effects. In this study, $\Delta\alpha$ is defined as the difference between reconstructed flip angle α_{AFI} and the real value α_{real}

$$\Delta\alpha = \alpha_{\mathrm{real}} - \alpha_{\mathrm{AFI}}. \tag{2.10}$$

The systematic error due to neglecting T_1 effects is not present in MTM. The second error component is due to statistical noise, present in both techniques, AFI and MTM. This is described by the propagation of noise in the input images to the reconstructed map. For the corresponding mapping error, the standard deviation ς of the reconstructed flip angle was used as a measure. As described below, the systematic and statistical errors of AFI and MTM were compared using optimized sets of scan parameters, i.e., using the combinations of TR and α_{nom} leading to optimal accuracy of the considered approaches.

The analytical framework of generalized AFI B_p mapping (cf. Eq. 2.6) allows for the estimation of the mapping error by the law of error propagation

$$\varsigma^2_{\mathrm{gAFI},\alpha} = \sum_{n=1}^{N} \left(\frac{\partial f}{\partial S_n} \varsigma_{\mathrm{in}} \right)^2. \tag{2.11}$$

The function f is given by Eq. 2.6, and ς_{in} denotes the standard deviation of the input signals S_n. The variance $\varsigma^2_{\mathrm{AFI},\alpha}$ of the standard AFI ($N = 2$) α-reconstruction is obtained from Eqs. 2.11 and 2.7 and reads

$$\varsigma^2_{\mathrm{AFI},\alpha} = \varsigma^2_{\mathrm{in}} \left[\frac{\left(1 - R^2\right)^2 \left(S_1^2 + S_2^2\right)}{S_1^4 \left[(R-r)^4 - (rR-1)^2 (R-r)^2\right]} \right] \frac{1}{M}. \tag{2.12}$$

Here, as discussed above, M denotes the number of AFI averages acquired and $R = \mathrm{TR}_2/\mathrm{TR}_1$. A corresponding expression for $N = 3$ was derived accordingly, and is shown in the appendix A.1.

The error propagation for the MTM fitting approach is described in the more general framework of the Cramér-Rao theorem (CRT) [50]. To this goal, it is first assumed that the measured signals s'_k ($k \in \{1 \ldots K\}$, $K = N \cdot M$ different TRs) can be written as $s'_k = s_k[\theta] + w_k$. White Gaussian noise is denoted by w_k, the unknown parameters by $\theta = [\alpha, T_1, S']$ and the signals $s_k[\theta]$ are given according to Eq. 2.1. It is further assumed that there is no cross correlation between noise in measurements and that the input variance of each measurement ς_{in} is independent of the object parameters θ. The Fisher information matrix $\mathbf{I}(\theta)$ required for CRT then reads [50]

$$[\mathbf{I}(\theta)]_{u,v} = \frac{1}{\varsigma_{\text{in}}^2} \sum_{k=1}^{K} \frac{\partial s_k[\theta]}{\partial \theta_u} \frac{\partial s_k[\theta]}{\partial \theta_v}. \tag{2.13}$$

According to CRT, the variance of an algorithm fitting the parameter set $\theta = [\alpha, T_1, S']$ possesses the lower bound

$$\varsigma_{\text{MTM},u}^2 \geq \left| \left[\mathbf{I}^{-1}(\theta) \right]_{u,u} \right| \tag{2.14}$$

where $\varsigma_{\text{MTM},u}^2$ denotes the variance of reconstructed $u \in \{\alpha, T_1, S'\}$. According to Eq. 2.14, the best settings for TR_{nm} and α_{nom} will depend on the parameter α, T_1 or S' chosen to be optimized. Since this work is mainly focused on B_p mapping, the optimal parameters that minimize $\varsigma_{\text{MTM},\alpha}^2$ have been chosen in most of the simulations and experiments. The minimization of $\varsigma_{\text{MTM},T_1}^2$ is investigated experimentally in Sections 2.4.2 and 2.4.3. For minimization, an absolute error function,

$$\delta_{\text{abs}} = (\varsigma_\alpha + \Delta\alpha) \sqrt{T_{\text{acq}}} \tag{2.15}$$

and a relative error function

$$\delta_{\text{rel}} = \frac{\delta_{\text{abs}}}{\alpha_{\text{real}}} \tag{2.16}$$

were defined, with ς_α being the corresponding standard deviations of AFI or MTM, respectively. $\Delta\alpha$ is given in Eq. 2.10, and thus, only nonzero in the case of AFI. The relative weight of ς_α and $\Delta\alpha$ in Eqs. 2.15 and 2.16 is influenced by the ratio $M_0/\varsigma_{\text{in}}$ of the input images. For minimizing

Eqs. 2.15 and 2.16, $M_0/\varsigma_{\text{in}} = 200$ was chosen in accordance with typical experimental findings. Noise differences caused by different total acquisition times (proportional to $T_{\text{acq}} = \sum_{n=1}^{N} \sum_{m=1}^{M} \text{TR}_{nm}$) are compensated in Eqs. 2.15 and 2.16 by the factor $\sqrt{T_{\text{acq}}}$. The described techniques allow the determination and comparison of the absolute and relative mapping errors of AFI and MTM B_p mapping as a function of the standard deviation ς_{in} of the underlying FFE images. The best settings for TR_{nm} and α_{nom} for MTM and AFI have been determined by minimizing the error functions Eqs. 2.15 and 2.16, respectively. The minimization was performed using in-house software written in C++, using the Downhill Simplex method [51].

In principle, the parameter space for comparing AFI and MTM is infinitely large. However, restrictions can be imposed for T_{acq} and for T_1. For the sake of a realistic total acquisition time, an upper limit for $T_{\text{acq}} < T_{\max} = K \cdot 125\text{ms}$ was chosen. The minimal TR_{nm} allowed for the Simplex method was set to 20 ms. On the other hand, only $T_1 < 1000$ ms was regarded, since larger T_1 values influenced optimization results only marginally, i.e., resulting TR values and flip angles varied less than 5 % for 1000 ms $< T_1 <$ 2000 ms and less than 10 % for 2000 ms $< T_1 <$ 3000 ms.

To start, standard $N = 2$ AFI was compared with generalized $N = 3$ AFI. Two optimal sets of parameters were obtained for each method by minimizing error functions δ_{abs} (Eq. 2.15) and δ_{rel} (Eq. 2.16), respectively. The first parameter set, obtained from minimizing δ_{abs}, was used to compare the quantities

$$\varsigma = \varsigma_\alpha \sqrt{T_{\text{acq}}} \tag{2.17}$$

and $\Delta\alpha$ (Eq. 2.10). The second parameter set, obtained from minimizing δ_{rel}, was used to compare

$$\text{SNR} = \frac{\alpha_{\text{real}}}{\varsigma} \tag{2.18}$$

Then, standard AFI and MTM were compared analogously. First, this was performed for $N = M = 2$ for the chosen T_1 range. A second comparison for $N = 2$ included increasing M, indicating multiple subsequent measurements in MTM and multiple averages in AFI. It was performed for distinct T_1 values and repeated for $N = 3$. All calculations were based on the assumption that noise on the input images ς_{in} is equal for both methods, and thus, cancels out by forming ratios (cf. Eqs. 2.12 and 2.14).

2.3.3 Experiments

All experiments were carried out on a 1.5 T Achieva system (Philips Healthcare, Best, The Netherlands). The quadrature body coil was used for RF transmission and a quadrature head coil for reception. An optimized RF spoiling scheme according to the approach presented in [43] was applied to each measurement m separately. For a discussion of MTM spoiling schemes please refer to Section 2.4.5.

2.3.3.1 Phantom study

Feasibility and accuracy A phantom consisting of six compartments with varying contrast agent concentrations resulting in different T_1 values was used. To obtain an independent reference to the MTM results, T_1 was additionally measured via the ratios and least squares algorithm (RLSQ), based on a combined spin echo (SE) / inversion recovery (IR) sequence [52], with $TR_{SE} = 1300\,ms$, $TR_{IR} = 3500\,ms$, TE spacing = 50 ms, and 8 echoes. An independent B_p reference was obtained using the double-angle method [30], with TR = 5 s and $\alpha_{1,2} = 60°/120°$.

As a first experiment, a three-dimensional volume of the phantom containing 30 slices (5 mm thickness) and field of view (FOV) of $220 \times 220\,mm^2$ with spatial scan resolution 96×98 pixels was acquired with TE = 1.72 ms. A dual TR sequence $N = 2$ with $M = 3$ different TR combinations (TR_{11} / TR_{21} = 50/200 ms, TR_{12} / TR_{22} = 50 / 250 ms, and TR_{13} / TR_{23} = 50 / 300 ms) was employed with a nominal flip angle $\alpha_{nom} = 50°$. Resulting imaging times were 12 min, 15 min, and 17 min, respectively. The B_p-related scan parameters α_{nom} and TR_{nm} were chosen to be in the optimal AFI range as given in [42]. Thus, the data have been reconstructed twice: via averaging analytical standard AFI reconstructions (Eq. 2.7) and via numerical MTM fitting (Eqs. 2.1 and 2.2).

For the six compartments, the systematic deviation between AFI B_p map and MTM B_p map has been determined as a function of the T_1 value measured by MTM, and compared with corresponding simulations. Moreover, the MTM T_1 map was compared with the RLSQ T_1 map and the MTM B_p map was compared with the double-angle B_p map.

Optimized for B_p mapping Using the same phantom, a second experiment was performed to analyze the noise sensitivity of AFI and MTM B_p

mapping and to compare the findings with theoretical predictions. Two dual TR pairs ($N = 2$, $M = 2$) were acquired. Simulations were performed employing error propagation theory according to the previous section to find optimal scanning parameters with respect to lowest noise in reconstructed B_p maps. Results of CRT noise minimization are presented in detail in Section 2.4.1. According to these results and target $T_1 = 375$ ms, a nominal flip angle $\alpha_{nom} = 88°$ and repetition times TR_{11} / $TR_{21} = 40$ / 146 ms and TR_{12} / $TR_{22} = 40$ / 473 ms were chosen for MTM mapping. Optimal performance for AFI field mapping was predicted for (TR_{11} / $TR_{21} = TR_{12}$ / $TR_{22} = 40$ / 309 ms) and $\alpha_{nom} = 58°$. All other sequence parameters were equal for both methods; 96×96 scan matrix, a FOV of $210 \times 210\,\text{mm}^2$, and 10 slices with 4 mm thickness. The same total imaging time of 11 min was achieved in both cases. Every sequence was run twice and corresponding reconstructed B_p images subtracted to determine the noise level of each method.

Optimized for T_1 mapping A third experiment was performed to analyze the MTM T_1 mapping performance and to compare with gold standard T_1 mapping. Experiments were conducted using a calibrated phantom (Test Object 5, Eurospin II Test System, Diagnostic Sonar LTD). Before the experiment, the temperature was measured inside the bore (see caption of Fig. 2.8 for resulting phantom manufacturer T_1 values). Sequence parameters for MTM were optimized with respect to maximum SNR in T_1 mapping using CRT. For optimization, a target $T_1 = 900$ ms was chosen. The minimum TR allowed in the parameter optimization was adjusted to 105 ms allow interleaved read-out of all slices. An in-plane resolution of $0.98 \times 0.98\,\text{mm}^2$ and 10 mm slice thickness (4 slices) was chosen. MTM T_1/B_p mapping was applied using $TR_{11,12} = 115/750$ ms, $TR_{21,22} = 105/202$ ms, and $TR_{31,32} = 105/130$ ms. An inversion recovery single shot TSE sequence (IR-TSE) (see Section 1.4.2) served as an independent T_1 measurement and performance reference (TR = 10 s, IR delays $t_{IR} = 5000/2244/1582/1188/906/686/506/353/221/105$ ms). For better comparison, acquisition times were adjusted to approximately 5 min for each sequence. The same sequences were also used for imaging a healthy volunteer.

2.3.3.2 Volunteer study

Experiments on healthy volunteers were conducted to proof the *in vivo* feasibility of the MTM approach. Written consent was obtained from all volunteers. Three-dimensional AFI images of the brain were acquired with coronal scan orientation, flip angle $\alpha_{\text{nom}} = 50°$, spatial resolution of $1.25 \times 1.25 \times 12\,\text{mm}^3$, and a FOV of $160 \times 160 \times 228\,\text{mm}^3$. Subsequent dual TR sequences ($N = 2$) were applied with $M = 3$ different TR combinations $\text{TR}_{11}/\text{TR}_{21} = 40/80\,\text{ms}$, TR_{12} / $\text{TR}_{22} = 40$ / $200\,\text{ms}$, and TR_{13} / $\text{TR}_{23} = 40$ / $400\,\text{ms}$, leading to imaging times of 4 min, 8 min, and 15 min. Repetition times and nominal flip angle represent optimal settings with respect to maximum SNR of B_p and T_1 mapping $\alpha_{\text{nom}}/\varsigma_{\text{MTM},\alpha} + T_1/\varsigma_{\text{MTM},T_1}$ for given total imaging time and $T_1 = 700\,\text{ms}$. TRs obtained via CRT have been rounded to yield integer ratios between TR pairs to allow spoiling of transverse magnetization according to [43]. In a post-processing step, the images were compensated for translatory and rotational patient motion between subsequent measurements, using a dedicated registration library [53].

2.4 Results

2.4.1 Performance analysis

Noise analysis of AFI and MTM B_p mapping employing error propagation theory is shown in Figs. 2.2-2.4. First, generalized $N = 3$ AFI was compared with standard $N = 2$ AFI and the results are shown in Fig. 2.2.

As mentioned, the ratios $\varsigma_{\text{AFI},N=3}/\varsigma_{\text{AFI},N=2}$ and $\Delta\alpha_{\text{AFI},N=3}/\Delta\alpha_{\text{AFI},N=2}$ are shown for minimized absolute error δ_{abs} (Eq. 2.15) and the ratio $\text{SNR}_{\text{AFI},N=3}/\text{SNR}_{\text{AFI},N=2}$ for minimized δ_{rel} (Eq. 2.16). Normalized SNR and standard deviation are shown according to definitions Eqs. 2.17 and 2.18. In the parameter range investigated, generalized AFI does not yield improvement over standard AFI with regard to SNR or standard deviation. However, generalized AFI has significantly lower systematic T_1 errors for $T_1 > 250\,\text{ms}$.

Second, the comparison of Fig. 2.2 was repeated for standard AFI and a dual TR sequence $N = 2$ and two subsequent measurements $M = 2$ (Fig. 2.3). For $T_1 \leq 400\,\text{ms}$, SNR_{MTM} is more than 30 % higher than SNR_{AFI},

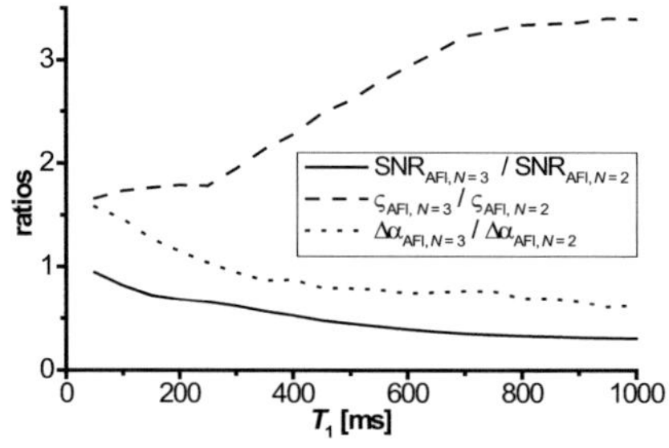

Figure 2.2: Comparison of standard $N = 2$ AFI and generalized $N = 3$ AFI. Using $N = 3$ different repetition times within one sequence is advantageous only with regard to lower systematic errors for $T_1 > 250\,\text{ms}$.

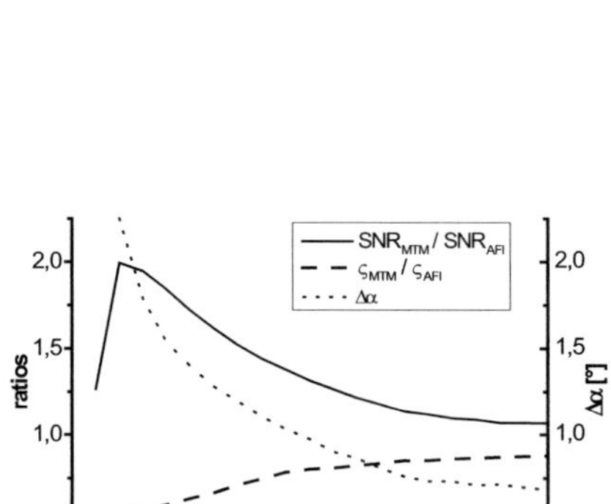

Figure 2.3: Comparison of $N = M = 2$ MTM and standard $N = 2$ AFI. The relative SNR increase is found to be between 1.1 and 2.0 at high T_1 (1 s) and low T_1 (100 ms), respectively. On the other hand, MTM is able to reduce noise by more than 10 % over a wide range of T_1 values.

	SNR		std deviation	
	$T_1 = 600\,\text{ms}$	$T_1 = 950\,\text{ms}$	$T_1 = 600\,\text{ms}$	$T_1 = 950\,\text{ms}$
TR_{11}	20 ms	20 ms	20 ms	20 ms
TR_{21}	105 ms	105 ms	55 ms	40 ms
TR_{12}	20 ms	20 ms	20 ms	20 ms
TR_{22}	355 ms	355 ms	405 ms	420 ms
α	118°	125°	82°	83°

Table 2.2: Optimal parameters found for MTM using CRT error propagation theory. All results shown are based on two subsequent dual-TR sequences $M = N = 2$ (see Fig. 2.3). Scan parameters for maximum SNR and minimal standard deviation in the reconstructed B_p maps, respectively, are shown for distinct T_1 values representing white ($T_1 = 600\,\text{ms}$) and gray ($T_1 = 950\,\text{ms}$) matter.

and standard deviation ς_{MTM} is more than 25 % lower than ς_{AFI}. For larger T_1, SNR gain by MTM decreases to roughly 7 % at $T_1 = 1000\,\text{ms}$. The highest systematic error found for AFI is $\Delta\alpha = 3.8°$ at $T_1 = 50\,\text{ms}$ and decreases with increasing T_1. The resulting optimal sequence parameters are discussed in Tab. 2.2 for distinct T_1 values.

To obtain maximum (normalized) SNR_{MTM}, nominal flip angles between 115° and 125° seem to be the best choice. For minimal variance, significantly lower optimal flip angles have been found (e.g., $\alpha_{\text{nom}} = 83°$ for $T_1 = 950\,\text{ms}$). The obtained optimal repetition times vary only slightly over the investigated T_1 range. As already observed for the optimal flip angles, optimizing SNR and noise variance yields different optimal TR_{nm}.

Third, more than $M = 2$ subsequent measurements have been analyzed and compared with averaging AFI results (Fig. 2.4). In the case of $N = 2$, SNR_{MTM} outperforms SNR_{AFI} for all investigated T_1 and M. In the case of $N = 3$, SNR_{MTM} is also higher than SNR_{AFI} (except for $M = 1$ and $T_1 \geq 950\,\text{ms}$) and similar to $\text{SNR}_{\text{MTM},N=2}$. The maximum SNR gain has been found for $M = 4(2)$ with $N = 2(3)$, which is caused by a trade-off between too small SNR for smaller M and too large T_{acq} for larger M. Since MTM requires at least $K = N \cdot M = 3$, the case $N = 2$ and $M = 1$ does not yield MTM results.

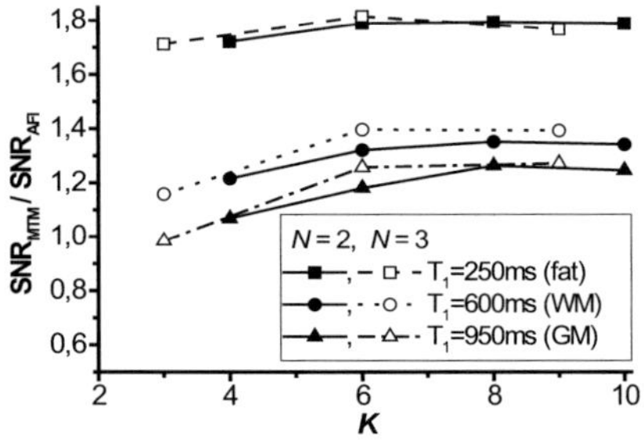

Figure 2.4: Comparing SNR of standard AFI and MTM using $N = 2$ (solid symbols) and $N = 3$ (open symbols). Ratios are shown as function of number of TRs $K = N \cdot M$ for three different T_1 settings representing gray ($T_1 = 950\,\text{ms}$) and white ($T_1 = 600\,\text{ms}$) matter as well as fat ($T_1 = 250\,\text{ms}$).

2.4.2 Phantom study

Experimental phantom results are shown in Figs. 2.5-2.10. Figures 2.5a and 2.5b show the measured flip angle maps of MTM and AFI. Both maps show a similar B_p distribution. The flip angle difference between the two methods is clearly T_1 dependent and increases with decreasing T_1 as discussed in connection with Fig. 2.2. Corresponding quantitative results are shown in Fig. 2.6a.

The comparison with simulations shows that the differences in reconstructed flip angles between AFI and MTM are due to the discussed systematic error in AFI B_p mapping, which is not present in MTM. Even for very large T_1, MTM does not exhibit such errors (see additional phantom experiment in appendix A.2). No significant systematic flip angle differences between long TR double-angle method and MTM B_p mapping were found. Predicted noise ratios via error propagation theory could be confirmed in the second phantom experiment, which employed optimized parameter settings. Noise reduction for the given T_1 range was predicted theoretically between 40 % and 50 % and found experimentally between 45 % and 55 % (Fig. 2.7).

The MTM T_1 map is shown in Fig. 2.5c. A quantitative comparison with the RLSQ measurement is given in Fig. 2.6b. MTM T_1 mapping is able to reproduce this reference measurement with sufficient accuracy. Fitting a straight line through the origin, $T_{1,\text{MTM}} = a \cdot T_{1,\text{RLSQ}}$, yields an inclination of $a = 0.96 \pm 0.02$. MTM T_1 mapping results from the third phan-

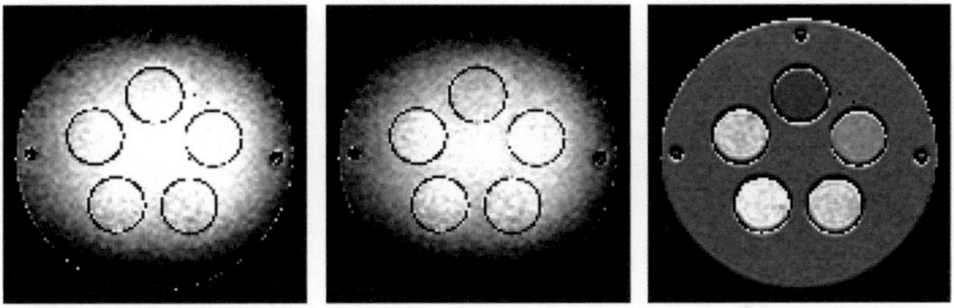

(a) (b) (c)

Figure 2.5: Flip angle map obtained with (a) MTM and (b) standard AFI reconstruction. Level and window were adjusted to highlight differences. The largest deviations can be observed in regions with low T_1, i.e., the background and the top tube, where AFI noticeably underestimates the actual flip angle. (c) T_1 map of the phantom, reconstructed with MTM. Quantitative values for B_p and T_1 are shown in Fig. 2.6.

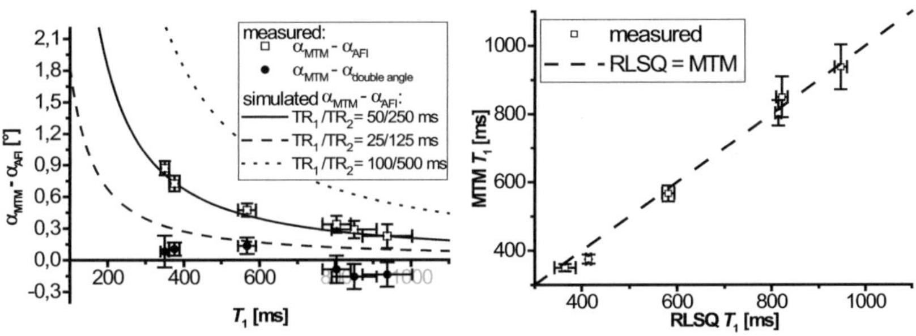

(a) (b)

Figure 2.6: Quantitative analysis of phantom experiments. (a) T_1 dependent difference in reconstructed flip angle between MTM and AFI (open symbols) and between MTM and the double angle method (solid symbols). The straight lines, representing the systematic error in AFI B_p mapping, were obtained from simulation of an AFI sequence with the same flip angle as observed in the experiment and different repetition times. The curves represent longer (dotted), same (solid), and shorter (dashed) TR_{nm} as used in the experiment. (b) shows T_1 values reconstructed with MTM as a function of T_1 values determined independently with a RLSQ sequence. Data points for both plots were obtained by averaging circular areas inside the tubes and on the background of the phantom (see Fig. 2.5).

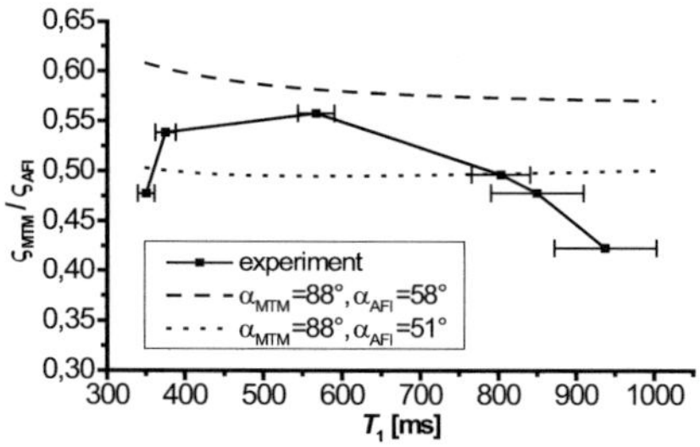

Figure 2.7: Comparison of standard deviation of MTM and AFI for $M = N = 2$. Noise reduction was predicted theoretically between 40 % and 50 % for flip angles α_{AFI} between 51° and 58°, as observed in the experiment. Noise ratios from experiments are found in good agreement to CRT calculations.

tom experiment using sequence parameters optimized for T_1 are shown in Figs. 2.9-2.10.

The IR-TSE T_1 map of the Eurospin phantom is shown in Fig. 2.8a. MTM T_1 and B_p maps are shown in Figs. 2.8b and 2.8c. A quantitative comparison is shown in Fig. 2.9. The fourth slice is not displayed since air bubbles inside the tubes spoiled average T_1 values. MTM T_1 phantom results were found in agreement with phantom manufacturer specifications. Disregarding the fourth slice, the maximum deviation from the true value was 126 ms for MTM and 257 ms for IR-TSE. In comparison with IR-TSE, MTM results were superimposed by stronger noise. Maximum noise (1/SNR) was 11.5 % for MTM and 1.4 % for IR-TSE. Predicted SNR via CRT could also be confirmed for MTM T_1 mapping (Fig. 2.10).

2.4.3 *In vivo* experiments

MTM results of proof of concept *in vivo* experiments are shown in Fig. 2.11. The B_p distribution with high resolution and low noise is shown in Fig. 2.11a. The T_1 distribution is shown in Fig. 2.11b, revealing good contrast between gray matter, white matter, and cerebrospinal fluid (CSF). The quantitative values are in good agreement with literature values [54]. The FFE image S_{nm} for $n = 1$, $m = 2$ is shown in Fig. 2.11c. Fig. 2.12 shows B_p and T_1 for optimized parameters for imaging the latter.

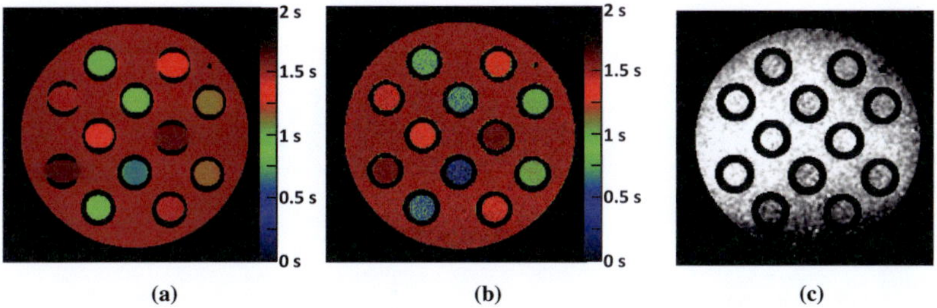

Figure 2.8: IR-TSE T_1 map (a) and MTM T_1 map (b). Simultaneously acquired MTM B_p map (c). T_1 values provided by manufacturer (left to right, top to bottom, T=23°C, units [ms]): 1278, 822, 664, 1385, 1137, 828, 496, 506, 1603, 1063, 1403, 659.

2.4.4 Steady-state investigation

To apply a quantitative imaging method as described in this chapter, it is crucial that the measured signal indeed follows the theoretical expectations, i.e., can always be described by the analytical expressions. Using steady-state sequences, it might be interesting whether large flip angle, long TR, or potential system imperfections will hamper the formation of a steady-state and lead to deviations from the analytical formula. To test steady-state formation, a bottle phantom was used and the measured signal (averaged over a small number of pixels inside the phantom) was compared to the expected (simulated) signal. Dual TR steady-state sequences with $N = 2$ were acquired with five different nominal flip angles $\alpha = 60/90/120/150/180°$. For each flip angle, the second repetition time was varied according to $\mathrm{TR}_{21} = 100/200/250/300/400\,\mathrm{ms}$ and the first TR was kept fixed at $\mathrm{TR}_{11} = 20\,\mathrm{ms}$. The results are plotted in Fig. 2.13. The simulated "real" signal was obtained from Eqs. 2.1 and 2.2 for independently determined $T_1 = 245\,\mathrm{ms}$ and α/α_{nom}. For flip angles $\alpha \leq 120°$ and all repetition times, the measured signal agrees very well with the simulated signal according to the analytical expression Eq. 2.4. In this parameter range, both methods MTM fitting as well as generalized AFI are expected to behave as expected. At $\alpha = 150°$ some deviation for the second signal S_{21} are observed and for $\alpha = 180°$ large errors in both signals are seen. Thus, the use of very large flip angles will lead to failure of both MTM and generalized AFI, and is not recommended.

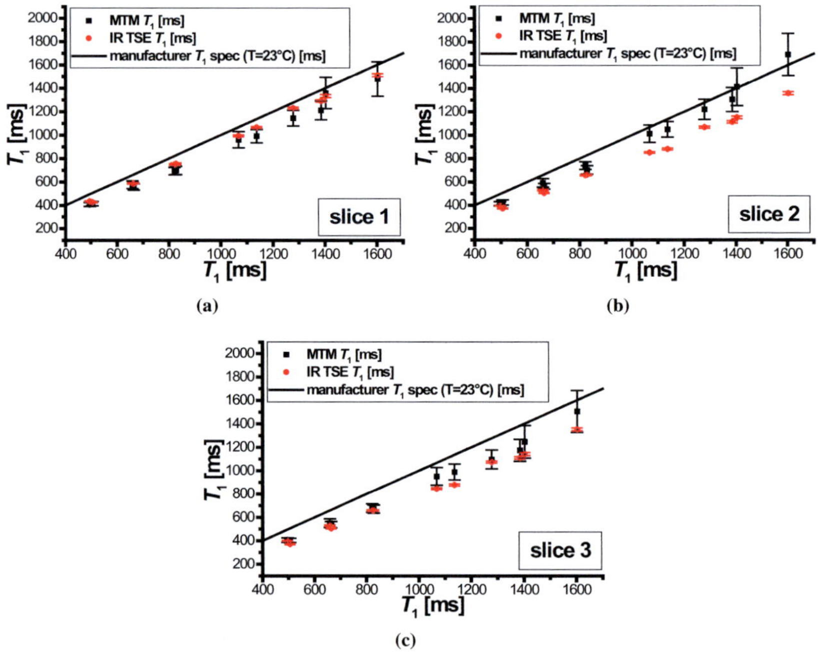

Figure 2.9: Comparison of T_1 results from IR-TSE sequence and MTM B_p T_1 mapping. MTM results are in good agreement with manufacturer specifications. The fourth slice is not shown since it did not yield reasonable quantitative values due to air bubbles inside the tubes.

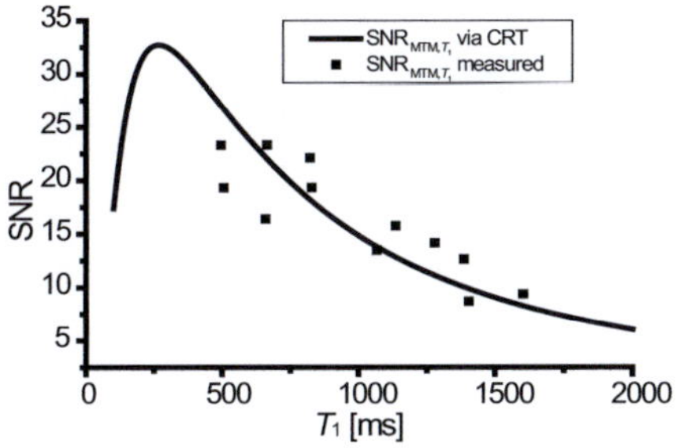

Figure 2.10: Comparison of predicted SNR using the Cramér-Rao theorem and measured SNR. SNR increases with decreasing T_1 and reaches approximately 20 for $T_1 \leq 1000$ ms. The plot is based on optimized parameters for T_1 mapping (see Section 2.3.3.1).

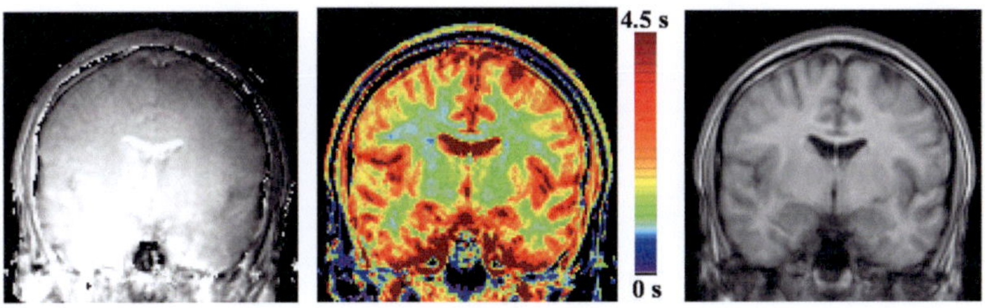

(a) (b) (c)

Figure 2.11: MTM *in vivo* results. B_p map (a), T_1 map (b), and corresponding anatomy (c) are shown. Flip angle maps with very low noise level are obtained (a). T_1 reveals good gray and white matter contrast (b,c). Total acquisition time was 27 min.

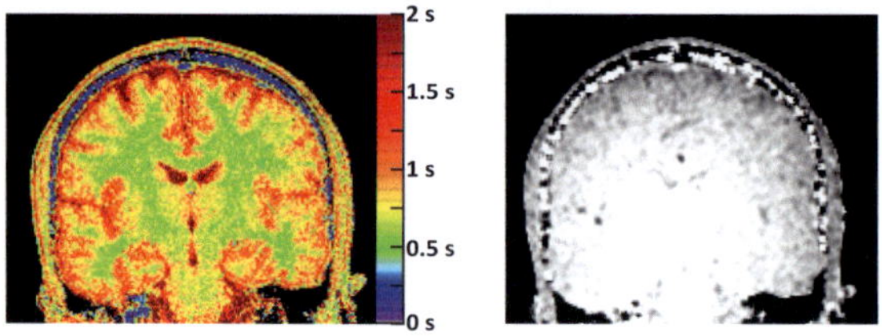

(a) (b)

Figure 2.12: MTM T_1 (a) and B_p (b) maps obtained in clinically acceptable total imaging time of 5 min. Parameters have been optimized for minimal variance in T_1 mapping.

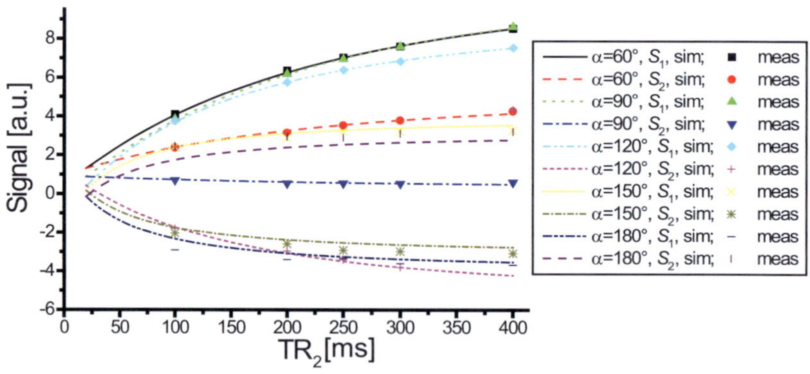

Figure 2.13: Comparison of measured and simulated signal as a function of TR_2 and flip angle α. A steady-state sequence with $N = 2$ and $TR_1 = 20$ ms was used. For flip angles $\alpha \leq 120°$ and all tested repetition times, negligible differences between expected steady-state signal and the measured signal are observed.

2.4.5 Optimized spoiling scheme for MTM

In standard AFI, spoiling of remaining transverse magnetization has been addressed in a number of publications [43, 44, 55, 56]. In this section, the approach presented in [43] has been used to study the effect of insufficient spoiling on the MTM reconstruction results. For two dual TR sequences ($M = 2$, $N = 2$) with repetition times $\mathrm{TR}_{11}/\mathrm{TR}_{21} = 30/210\,\mathrm{ms}$ and $\mathrm{TR}_{12}/\mathrm{TR}_{22} = 30/300\,\mathrm{ms}$ and flip angle $\alpha = 60°$ the signals S_{nm} were computed taking into account RF spoiling and any additional remaining transverse magnetization. According to the theory presented in [43], different values for RF spoiling pulse phase shifts Φ_m have been applied, where m denotes the first or second set of repetition times $m \in \{1,2\}$. The signals generated in this way have been used as input to the MTM fitting algorithm. The difference between reconstructed and expected flip angle is depicted in Fig. 2.14. Fig 2.14a is based on moderate diffusion damping $d = 0.2$, intermediate $T_1/T_2 = 800/50\,\mathrm{ms}$, and typical flip angle $\alpha = 60°$. In Figs. 2.14b to 2.14d, deviations from these standard settings are investigated. In Fig. 2.14b, long relaxation times $T_1/T_2 = 4500/2200\,\mathrm{ms}$ have been analyzed, Fig. 2.14c shows the error for low diffusion damping regime $d = 0.02$, and Fig. 2.14d takes smaller flip angles into account. Opposite deviations, i.e. lower T_1, higher diffusion damping, and higher flip angles all led to even smaller errors. The figures underline that many combinations of spoil phases yield minimal error over the whole range of physiological parameters. For the experiments described in this chapter, a Downhill Simplex algorithm [51] was implemented to obtain optimal values for spoil phases yielding minimal flip angle mapping error.

2.5 Discussion

The B_p reconstruction of the previously published technique AFI is based on two data points acquired per voxel, corresponding to the two different TRs forming the AFI sequence. The presented study raises the number of data points per voxel by introducing multiple TR sets and / or multiple sequences with differing TR sets. From the obtained data points, B_p can be reconstructed in two ways. On one hand, an analytical formula can be applied, which is derived from ratios of the obtained signals, in analogy to the standard AFI technique. On the other hand, the data points can be

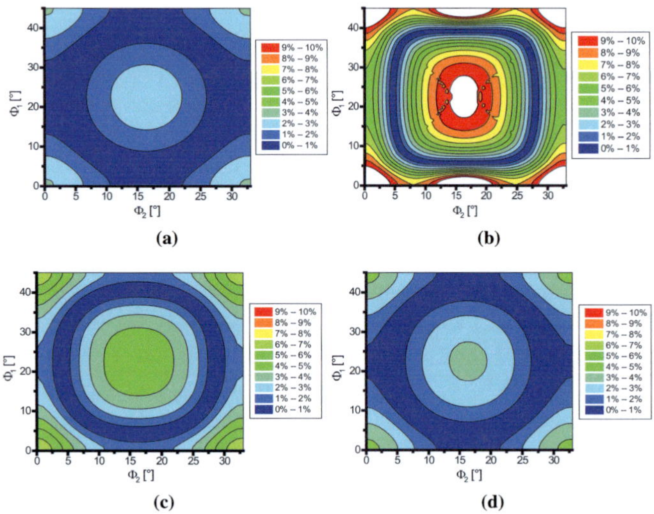

Figure 2.14: MTM B_p mapping error in % due to incomplete spoiling of transverse magnetization for different phase increments. Φ_1 is the spoiling phase used in the first of $M = 2$ TR pairs, Φ_2 in the second. Although both phases in general range from 0 to 360°, the choice of TRs leads to periodic error patterns [43] and only one period is displayed. Fig. (a) uses $d = 0.2$, $T_1/T_2 = 800/50\,\text{ms}$, and $\alpha = 60°$. This parameter set has been varied by (b) changing relaxation times to $T_1/T_2 = 4500/2200\,\text{ms}$, (c) changing diffusion damping to $d = 0.02$, and (d) changing the flip angle $\alpha = 50°$. A wide range of possible optimal combinations of spoil phases yielding minimum error is found for all investigated cases.

fitted in the framework of a numerical model. In opposite to the analytical procedure, this technique does not only yield B_p, but also T_1.

The investigated extension of the analytical approach from $N = 2$ to $N = 3$ ("generalized AFI", Fig. 2.2) exhibited a significant drop of the systematic B_p error, which is caused by the assumption $T_1 \gg \mathrm{TR}$. However, this benefit is counterbalanced by a strong noise increase. Thus, in the framework of the analytical approach, no clear advantage of $N = 3$ is visible.

Alternatively, the data points obtained with the discussed $N = 3$ sequence can serve as input for the numerical model fitting ("MTM", Fig. 2.3). This reconstruction method shows a more benign error propagation and a higher SNR than standard AFI, at least for a wide range of T_1. Using more than one TR triple (i.e., $M \geq 2$, $N = 3$), this advantage increases, and holds for all investigated T_1. Applying the same acquisition times, the SNR gain for $N = 3$ is similar to $N = 2$ (keeping in mind that $N = 2$ is possible only for $K \geq 4$). Again, a separate study shall analyze if this observation also holds for higher N.

In the performance analysis $M_0/\varsigma_{in} = 200$ was used, determining the relative weight of systematic and statistical error in the AFI optimization. This factor is related to but not identical with image SNR. To translate the factor M_0/ς_{in} into SNR, repetition times, flip angle, and T_1 have to be taken into account. In the experiments conducted in this study, the image SNR ranged from 8 % to 40 % of M_0/ς_{in}. The obtained results advise that the SNR of AFI should not be increased by simply averaging multiple measurements, i.e., repeating the AFI sequence with identical TR pairs. Instead, available scan time ought to be used for sampling M different TR sets, enabling B_p reconstruction via MTM fitting. For a given N, an optimum M can be determined, yielding maximum SNR gain compared with straight-forward averaging. In general, this optimum depends on the choice of the maximum summed profile duration $T_{\max}$, set to $K \cdot 125\,\mathrm{ms}$ in this study, yielding $M = 4(2)$ as optimum for $N = 2(3)$. This advantage could also be used to accelerate B_p mapping maintaining SNR.

Although presented investigations employed quadrature coils exclusively, the combination of MTM with parallel imaging (see, e.g. [57]) seems to be particularly advantageous. At least at low acceleration factors A, the SNR loss of parallel imaging shows the statistical behavior proportional to $\sqrt{A}$. According to the presented study, MTM is able to over-compensate this SNR loss. Of course, imaging time in general will depend on the readout mode used, e.g. EPI and/or half Fourier techniques. While predicted SNR gain is expected for all T_1 values investigated, it is largest at $T_1 \approx 150\,\mathrm{ms}$, where, however, no corresponding tissue types have been reported for main field strength $B_0 = 1.5\,\mathrm{T}$ used in the presented experiments [54]. Here, it might be worth to mention that B_0 enters MTM exclusively via its impact on T_1, i.e., the presented calculations hold for all B_0 as long as adequate T_1 values are considered. For low-field systems $B_0 < 0.5\,\mathrm{T}$, also T_1 values in the above mentioned range of 150 ms occur.

Strongly varying B_p fields, as can occur in parallel transmission [28] [29], are frequently problematic for B_p mapping techniques. This aspect has not been investigated in this study. However, it can be circumvented by the recently introduced "all-but-one-channel" approach [58, 59], yielding B_p maps with significantly reduced dynamic range.

In the experiments conducted in this study, $\mathrm{TR}_{nm} \leq T_1$ has been used. Considering even smaller TR_{nm} in the case of AFI B_p mapping would lead to lower systematic errors on the one hand and an increase of noise on the other hand. In the case of MTM, smaller TR_{nm} would lead to an increase of noise level for both B_p and T_1 maps. The relative noise increase would depend on the optimization scheme chosen, i.e. minimizing $\varsigma^2_{\mathrm{MTM},T_1}$ or $\varsigma^2_{\mathrm{MTM},\alpha}$. To take into account this noise dependence on imaging time, SNR and variance have been normalized with imaging time in the numerical comparison presented in this study (Eq. 2.15).

MTM simultaneously provides a T_1 map with good accuracy without requiring additional acquisition time. Thus, MTM surpasses subsequent B_p and T_1 measurement techniques, which are often performed to correct one measurement by use of the other. Furthermore, the B_p transmit field determined with MTM is intrinsically T_1 corrected, and thus, has no systematic error as found in AFI. An investigation of MTM's T_1 mapping efficiency and comparison to a single shot implementation of an IR-TSE T_1 mapping sequence yielded comparable or even better accuracy, however, at the cost of worse SNR. MTM could be combined with EPI read-out to increase SNR of T_1 and B_p mapping. The simultaneous nature of acquisition in MTM is expected to be particular suitable in T_1 mapping at large excitation field variations.

With regard to incomplete spoiling of remaining transverse magnetization, it is assumed that different measurements with different settings for repetition times will result in varying contributions of spurious spin echoes to the signal [43]. A fit algorithm could possibly eliminate these effects, and therefore, could be advantageous over systematic approaches. Thus, MTM is less prone to systematic errors from spurious transverse magnetization. It could be confirmed that errors in MTM stemming from remaining transverse magnetization behave rather benignly. A single setting for RF spoiling phases yields good B_p mapping accuracy over a wide range of parameters influencing spoiling (diffusion damping, flip angle, T_1). Regarding AFI, in case of incomplete spoiling, Eq. 2.6 remains only valid if the ratio of the signals S_i and S_j is preserved. Otherwise, systematic deviations will potentially result in biased flip angle maps. It seems that a more careful spoiling scheme than for the MTM fitting approach should be used, since optimal spoiling settings depend heavily on applied object and scan parameters. Besides remaining transverse magnetization, other problems predominantly encountered in *in vivo* imaging have not been addressed in the presented study. Flow could be a source of error in B_p mapping results, particularly, CSF flow might account for altered B_p visible in Fig. 2.11a (see also appendix A.2).

In principle, generalizing the AFI approach could also involve varying flip angles between different segments. In this case, however, a more sophisticated optimization scheme has to be used. An investigation is therefore postponed to a separate study.

Only the $N = 2$ steady-state has been implemented on the scanner, since more images per steady-state sequence would have required significant amount of programming work. Especially $N = 3$ seems to be promising because MTM fitting could be based on the minimum number of required data points acquired within a single sequence.

2.6 Conclusion

A new simultaneous B_p / T_1 mapping approach named MTM has been introduced. Its B_p mapping performance was thoroughly compared with AFI, a well established, fast B_p mapping technique. The new approach is able to substantially decrease systematic and statistical mapping errors compared to standard AFI, which is beneficial for various applications that rely on accurate and fast B_p mapping. T_1 mapping capabilities of MTM were tested and revealed high degree of accuracy, although speed and SNR still leave room for improvement compared to highly optimized reference sequences. The applied Cramér-Rao lower bound estimation allows a fast determination of optimal scanning parameters for performing MTM sequences. *In vivo* applications of MTM shown in this study were limited to imaging the head. In principle, this approach is applicable for all anatomies, which will be subject of further investigations. Applications of MTM B_p/T_1 mapping seem to be especially beneficial in electric properties tomography (see next chapter), which heavily relies on accurate B_p mapping, as well as perfusion imaging based on T_1 DCE-MRI (see Section 1.4.5), where B_p inhomogeneity induced errors are a major obstacle in baseline T_1 mapping. Hence, perfusion imaging could benefit substantially from a simultaneous measurement of both parameters.

2.7 Acknowledgments

The authors of [60] thank Ewald Rössl for valuable contributions to the work on Cramér-Rao lower bound calculations and Christian Stehning for help on motion correction of volunteer data.

CHAPTER THREE

ELECTRIC PROPERTIES TOMOGRAPHY: IMAGING DIELECTRIC PROPERTIES USING MRI

Wär nicht das Auge sonnenhaft, die Sonne könnt es nie erblicken.

— JOHANN WOLFGANG VON GOETHE (1749 - 1832)

Abstract — The electric properties of human tissue can potentially be used as a diagnostic parameter, e.g., in tumor diagnosis. In the framework of radiofrequency (RF) safety, the electric conductivity of tissue is needed to correctly estimate the local specific absorption rate (SAR) distribution during mobile phone communication and MR measurements. In this study, a recently developed approach, called electric properties tomography (EPT), is adapted for and applied to *in vivo* imaging. It derives the patient's electric conductivity and permittivity from the spatial sensitivity distributions of the applied RF coils. In contrast to other methods to measure the patient's electric properties, EPT does not apply externally mounted electrodes, currents, or RF probes, which enhances the practicability of the approach. This work shows that conductivity distributions can be estimated from phase images and permittivity distributions can be estimated from magnitude images of the RF transmit field. Corresponding numerical simulations using finite-difference time-domain methods support the feasibility of this phase-based conductivity imaging and magnitude-based permittivity imaging. Using this approximation, three-dimensional *in vivo* conductivity and permittivity maps of the human brain are obtained in 5 and 13 min, respectively, which can be considered as a step towards clinical feasibility of EPT.

Based upon:

- T. Voigt, U. Katscher, O. Dössel; Quantitative Conductivity and Permittivity Imaging of the Human Brain using Electric Properties Tomography. *Magnetic Resonance in Medicine* 2010, *in press*.
- T. Voigt, U. Katscher, O.Doessel; In vivo Quantitative Conductivity Imaging Based on B1 Phase Information. In *Proceedings of the 18th Annual Meeting of ISMRM*, Stockholm, Sweden 2010. p. 2865.

3.1 Introduction

Magnetic Resonance Tomography provides a vast variety of possible image contrasts. Due to reasons of reproducibility and comparability, contrasts comprising quantitative parameters are of particular clinical interest. Current examples of quantitative MRI techniques are diffusion, perfusion, and permeability imaging; however, also electric conductivity and permittivity are possible candidates for quantitative parameters. The idea of extracting these electric properties from MR images was already proposed in 1991 [61]. However, only recently, the electric properties of the human body have been introduced as a quantitative image contrast in standard MRI via electric properties tomography (EPT) [62]. EPT allows the determination of the conductivity and permittivity using the radiofrequency (RF) transmit field map of a standard MR scan [63].

The task of imaging electric properties has been addressed by a variety of imaging modalities. Among these modalities, electric impedance tomography (EIT) is probably the most prominent one. It is performed by applying low frequency currents through multiple electrodes and reconstructing electric properties by solving the resulting inverse problem [64–67]. Magnetic induction tomography (MIT) is a similar approach, however, using RF coils for current induction and reception of the resulting fields [68, 69]. In Magnetoacoustic tomography with magnetic induction (MAT-MI), acoustic pressure is generated by applying a time varying magnetic field to a sample inside a static magnetic field. Emitted ultrasonic waves are detected and used for reconstructing electric properties. [70,71]. Magnetic resonance electric impedance tomography (MREIT) is a method initially based upon EIT, including electrode mounting, however, taking advantage of the spatial encoding inherent to MRI [72–74]. RF current density imaging (RF-CDI) makes electrode mounting obsolete by the application of a separate RF pulse for current induction, again taking advantage of MRI spatial encoding [75]. Another MR based method estimates conductivity assuming proportionality between conductivity and the diffusion coefficient of water [76,77]. MR noise tomography aims at extracting dielectric properties from noise of MR images [78, 79]. All these methods differ from the presented approach, which uses a standard MR system and requires neither electrode mounting nor the application of additional

	Conductivity	Permittivity
Colon	1.19	1.25
Kidney	0.97	0.94
Liver	1.35	1.06
Lung	1.32	0.90
Mammary	7.09	3.37

Table 3.1: Ratios of dielectric properties of malignant to healthy tissue acquired *ex vivo* at 100 MHz. Especially for breast tumors but also in liver, lung, and colon tumors, dielectric properties are of diagnostic value. Data has been extracted from [82].

RF energy. Instead, the EPT approach employs post-processing the field map of the imaging RF pulse. Since electrical properties are in general frequency dependent, EPT yields quantitative values for conductivity and permittivity at MR Larmor Frequency.

Ex vivo studies suggest diagnostic value of electric properties in the field of, e.g., heart muscle viability after myocardial infarction [80]. Significant changes of conductivity and permittivity have also been reported in *ex vivo* studies on brain tissue in connection with stroke [81]. In the field of oncology, studies suggest that tumors can be characterized by their electrical properties, distinguishing between healthy and malignant tissue (see, e.g., [82] and Tab. 3.1). A study focused on dielectric properties of breast carcinoma revealed strong differences in *ex vivo* conductivity of tissue containing tumor cells and healthy tissue [83]. To summarize the *ex vivo* findings, dielectric properties are expected to yield useful information on the underlying physiology and pathology that might be beneficial in diagnostic as well as therapeutic decisions.

On the other hand, electric properties are not only of potential diagnostic value, but also central parameters in the field of RF safety. Here, the local heating of tissue is a major problem at high field MR, particularly in the framework of parallel transmission [84–86]. The acceptable local specific absorption rate (SAR), which is directly related to tissue heating, may limit the parameter space available for the application of specific MR sequences. The typically used finite-difference time-domain (FDTD) simulations to determine local SAR require a dielectric patient model as input. It has been shown that SAR simulations based on non-specific dielectric patient models suffer from limited accuracy [87]. Thus, an exact determination of local SAR has to be based on patient specific data. The EPT approach presented in this chapter provides a method for patient specific imaging of dielectric properties, which can be used in local SAR caculations (see Chapter 4).

EPT and similar approaches have been applied for simulation studies [88–90], for phantom studies using quadrature excitation [91], and using parallel transmission [92–94]. In this study, EPT is applied to *in vivo* cases [95, 96]. Quantitative conductivity and permittivity values of the human brain are measured using a standard MR system. Based on EPT, phase-based conductivity imaging and magnitude-based permittivity imaging are introduced. Both techniques have been developed in the framework of this study. Thus, it is the first time that they are tested in simulations and applied to *in vivo* experiments on healthy volunteers.

3.2 Sensitivity of MRI fields to dielectric properties and physiology

The dielectric properties of the sample have an impact on the images created in MRI. According to Faradays law, any changing magnetic flux will induce eddy currents. According to Lenz's law, the effect of these eddy currents will oppose the effect of incident RF. The electro-magnetic (EM) field used to flip the spins is time-varying with frequency ω_0 (cf. Section 1.3.1) and hence will be distorted by emerging eddy currents. As a result, the flip angle will be lower than the demand value in some parts of the image, and - due to superposition effects - higher in other parts of the image. In standard imaging, this will result in spatially dependent image intensity, perceived as reduced image quality. This effect also gives rise to errors in quantitative imaging techniques presented in Section 1.4. In EPT, these influences are measured to obtain information on the dielectric properties.

Figure 3.1 gives an overview on the impact of dielectric properties on the fields involved in MRI, based on the example of a homogeneous sphere of 10 cm radius (Fig. 3.2a). The impact is expressed as correlation between actual and vacuum field amplitude as well as root mean square difference (RMSD) between actual and vacuum phase, expressed in radians. Apart from some smaller structures, for all human tissues significant changes of amplitude and/or phase are observed and hence should be able to be

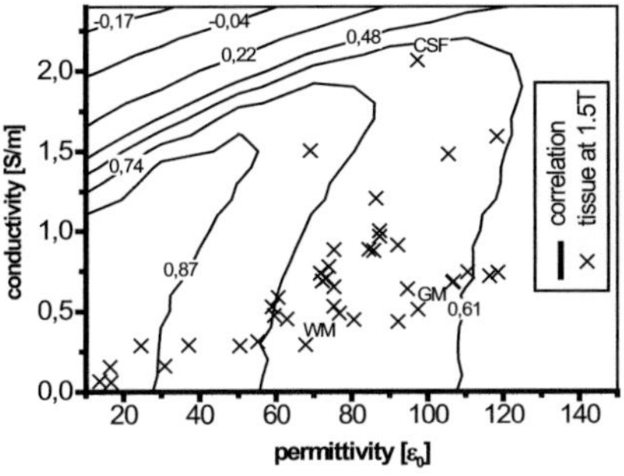
(a)

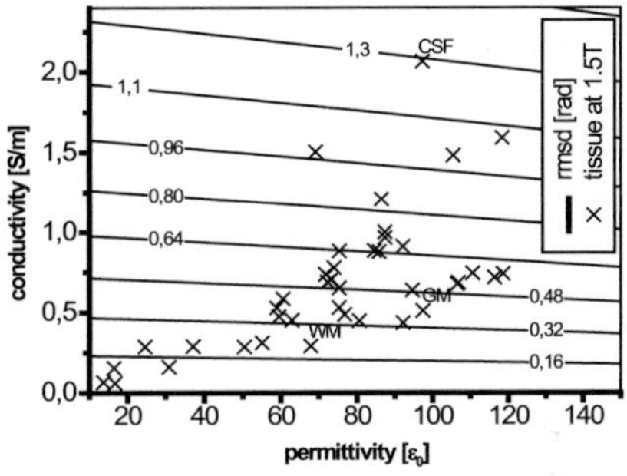
(b)

Figure 3.1: Dielectric properties of the sample distort the EM fields used in MRI. In (a), the influence of the dielectric properties on the B_p magnitude is shown. The correlation between actual and vacuum field is shown for the range of physiologic conductivity and permittivity values. In (b), the influence on the RF phase is shown, i.e., the root mean square difference between actual and vacuum phase, expressed in radians. Human tissue types are indicated by crosses [97]. .

recovered from EPT reconstruction.

The influence of the dielectric properties on the magnitude of the excitation field can be measured using, e.g., MTM (Chapter 2) or Bloch Siegert mapping (Section 1.4.1). The influence of the dielectric properties on the phase of the excitation field can be measured using, e.g., the image phase of a spin echo image (Chapter 3.3.4).

The macroscopic effects on length scales of millimeters detectable with MRI originate from microscopic properties of cell constituents. Thus, ideally the dielectric properties allow assessing underlying physiologic parameters. In humans, dielectric properties are determined at the cellular level. They are related to several factor such as ion concentration, molecular compounds of the cells, cells structure and density, membrane characteristics and other factors. The interplay of these factors leads to the frequency dependence of human tissue's dielectric properties observed in many *ex vivo* experiments (see e.g. [98, 99]). Usually, different frequency regimes of dispersion are distinguishable in conductivity/permittivity spectra [100], reflecting specific sub-cellular features. In particular, MR Larmor frequencies cover the crossover from so called β-dispersion to γ-dispersion regime. In this regime, membrane influence becomes small compared with the internal impedance and dispersion is not yet completely governed by polar properties of water [100, 101]. Human tissues at these frequencies can be considered as suspensions of non-conducting proteins and other solids in electrolyte [99]. An alteration of the effectual impact on the composition of intracellular cytoplasm and hence internal impedance should therefore be detectable by quantitative values in MR dielectric properties imaging. EPT should yield direct measurement of the cell interior, e.g. salt, water, or protein content and facilitate the study of metabolic processes.

3.3 Theory

3.3.1 Derivation of the central equations

EPT relates the patient's electric properties to quantities that are accessible via MRI. The transmitted RF pulse is affected by the conductivity and permittivity of the tissue being imaged. [63] formulates this relation by a suitable arrangement of Maxwell's Equations. First, Ampere's law is integrated along a closed path∂A that delineates the border of the area A

$$(i\omega)^{-1} \oint_{\partial A} \nabla \times \mathbf{H}(\mathbf{r})\,\mathrm{d}\mathbf{l} = \oint_{\partial A} \kappa(\mathbf{r})\,\mathbf{E}(\mathbf{r})\,\mathrm{d}\mathbf{l} \tag{3.1}$$

Here,ω is the Larmor frequency, and $\kappa = \varepsilon - {i\sigma}/{\omega}$ denotes the complex permittivity with ε being the scalar permittivity and σ the electric conductivity, and $\mathrm{d}\mathbf{l}$ is a line element. For the following derivations, κ is assumed to be isotropic. The magnetizing field[1] strength vector $\mathbf{H}$ and the electric field vector $\mathbf{E}$ correspond to the RF fields of the MR system, i.e., they can be assumed to be time-harmonic $\mathbf{H},\mathbf{E} \sim \exp(i\omega t)$. Then, Faraday's law reads

$$-i\omega\mu \int_{A} \mathbf{H}(\mathbf{r})\mathrm{d}\mathbf{a} = \oint_{\partial A} \mathbf{E}(\mathbf{r})\,\mathrm{d}\mathbf{l} \tag{3.2}$$

with μ being the (assumed to be constant) permeability and surface element $\mathrm{d}\mathbf{a}$. Dividing Eq. 3.1 by Eq. 3.2 relates the electric properties to the

[1] **H** will be referred to as magnetizing field and **B** as magnetic field according to the notation of [102]. Since constant μ is assumed, **H** and **B** differ only by a constant scaling factor.

magnetizing fields

$$\frac{\oint_{\partial A} \nabla \times \mathbf{H}(\mathbf{r})\,\mathbf{dl}}{\omega^2 \mu \int_A \mathbf{H}(\mathbf{r})\mathbf{da}} = \frac{\oint_{\partial A} \kappa(\mathbf{r})\,\mathbf{E}(\mathbf{r})\,\mathbf{dl}}{\oint_{\partial A} \mathbf{E}(\mathbf{r})\,\mathbf{dl}} \approx \kappa(\mathbf{r}) \tag{3.3}$$

Eq. 3.3 represent the EPT approach as used in [63]. The approximation used in the last step of Eq. 3.3 is valid in regions where the spatial variation of κ along ∂A is significantly smaller than the variation of E, e.g., if ∂A lies inside compartments with constant κ. Using Stokes theorem and the vector identity $\nabla \times \nabla \times \mathbf{H} = \nabla(\nabla \cdot \mathbf{H}) - \triangle \mathbf{H} = -\triangle \mathbf{H}$ yields

$$\frac{\int_A \nabla \times \nabla \times \mathbf{H}(\mathbf{r})\,\mathbf{da}}{\omega^2 \mu \int_A \mathbf{H}(\mathbf{r})\mathbf{da}} = \frac{-\int_A \triangle \mathbf{H}(\mathbf{r})\,\mathbf{da}}{\omega^2 \mu \int_A \mathbf{H}(\mathbf{r})\mathbf{da}} \approx \kappa(\mathbf{r})\,. \tag{3.4}$$

The following derivations leading to Eq. 3.8 employ positive and negative circularly polarized coordinates, indexed by p and m, respectively, and a z-component. The new basis vectors $\mathbf{e}_p$, $\mathbf{e}_m$, $\mathbf{e}_z$ are related to Cartesian $\mathbf{e}_x$, $\mathbf{e}_y$, $\mathbf{e}_{z,cart}$ via $\mathbf{e}_p = \frac{1}{\sqrt{2}}(\mathbf{e}_x + i\mathbf{e}_y)$, $\mathbf{e}_m = \frac{1}{\sqrt{2}}(\mathbf{e}_x - i\mathbf{e}_y)$, $\mathbf{e}_z = \mathbf{e}_{z,cart}$ and thus, are mutually orthogonal. The transformation $\Phi : \mathbb{C}^3 \to \mathbb{C}^3$ from Cartesian $\mathbf{r} = (x,y,z)^T$ to the circular coordinate system $\tilde{\mathbf{r}} = (p,m,z)^T$ can be described by the matrix

$$\mathbf{J} = \begin{pmatrix} \frac{1}{\sqrt{2}} & \frac{i}{\sqrt{2}} & 0 \\ \frac{1}{\sqrt{2}} & \frac{-i}{\sqrt{2}} & 0 \\ 0 & 0 & 1 \end{pmatrix}. \tag{3.5}$$

Using this matrix, the coordinate transformation and the inverse relation can be written as

$$\tilde{\mathbf{r}} = \mathbf{J} \cdot \mathbf{r} = \begin{pmatrix} \frac{1}{\sqrt{2}}(x+iy) \\ \frac{1}{\sqrt{2}}(x-iy) \\ z \end{pmatrix} \quad \text{and} \quad \mathbf{r} = \mathbf{J}^\dagger \cdot \tilde{\mathbf{r}} = \begin{pmatrix} \frac{1}{\sqrt{2}}(p+m) \\ \frac{i}{\sqrt{2}}(m-ip) \\ z \end{pmatrix} \tag{3.6}$$

with $\dagger$ denoting the Hermitian conjugate. The transformation matrix $\mathbf{J}$ is unitary $\mathbf{J}^\dagger \cdot \mathbf{J} = \mathbf{J} \cdot \mathbf{J}^\dagger = \mathbf{1}$ and identical with the Jacobian $\mathbf{J} = \frac{\partial \tilde{\mathbf{r}}}{\partial \mathbf{r}}$. In these circular coordinates, the magnetizing field vector reads $\mathbf{H} = (H_p, H_m, H_z)^T$ with $H_p = \frac{1}{\sqrt{2}}(H_x + iH_y)$, $H_m = \frac{1}{\sqrt{2}}(H_x - iH_y)$, and $H_z = H_{z,cart}$.

All three components of this magnetizing field vector enter Eq. 3.4; however, only H_p couples to the proton spins, and thus, is accessible via

MR. In the following, Eq. 3.4 is re-formatted to eliminate the inaccessible H_m and H_z. Since A is an arbitrarily orientated surface inside the object that is imaged, this additional degree of freedom can be used to further reduce the required input quantities. Using a surface element $\mathrm{d}\mathbf{a} = \mathbf{e}_p \mathrm{d}m\mathrm{d}z$ with its normal parallel to $\mathbf{e}_p$ substitutes $\mathbf{H}$ by H_p in the numerator and denominator of Eq.3.4. To this goal, the transformation Φ is applied to the surface integrals and the Laplacian operator. The surface integral for a surface element $\mathrm{d}\mathbf{a} = \mathbf{e}_p \mathrm{d}m\mathrm{d}z$ transforms according to

$$\begin{aligned} \mathbf{H}\mathrm{d}\mathbf{a} &= \begin{pmatrix} H_x \\ H_y \\ H_z \end{pmatrix} \cdot \mathbf{n}\mathrm{d}a = \begin{pmatrix} H_x \\ H_y \\ H_z \end{pmatrix} (\mathbf{r}_m \times \mathbf{r}_z)\, \mathrm{d}m\mathrm{d}z \\ &= \begin{pmatrix} H_x \\ H_y \\ H_z \end{pmatrix} \begin{pmatrix} \frac{i}{\sqrt{2}} \\ \frac{-1}{\sqrt{2}} \\ 0 \end{pmatrix} \mathrm{d}m\mathrm{d}z = -iH_p \mathrm{d}m\mathrm{d}z \end{aligned} \tag{3.7}$$

with $\mathbf{r}_m = \frac{\partial \mathbf{r}}{\partial m} = \begin{pmatrix} \frac{1}{\sqrt{2}} \\ \frac{i}{\sqrt{2}} \\ 0 \end{pmatrix}$, $\mathbf{r}_z = \frac{\partial \mathbf{r}}{\partial z} = \begin{pmatrix} 0 \\ 0 \\ 1 \end{pmatrix}$, and $\mathbf{n}$ being the normal of the surface. Hence, an additional factor of $-i$ is introduced by this coordinate transformation, which however cancels out by forming the ratio in Eq. 3.4. The Laplacian remains unchanged because unitary transformations do not affect scalar products such as $\tilde{\triangle} = \mathbf{J}\nabla \cdot \mathbf{J}\nabla = \mathbf{J} \cdot \mathbf{J}^\dagger \triangle = \triangle$.

An additional integration of Eq. 3.4 (where Eq. 3.7 has been applied) using line element $\mathrm{d}p$ along the direction of the positively circulating component $\mathbf{e}_p$ yields a volume integral

$$\frac{-\int_{\Phi^{-1}(l)} \left[\int_{\Phi^{-1}(A)} \triangle H_p(\Phi(\mathbf{r}))\, i\mathrm{d}m\mathrm{d}z \right] \mathrm{d}p}{\omega^2 \mu \int_{\Phi^{-1}(l)} \left[\int_{\Phi^{-1}(A)} H_p(\Phi(\mathbf{r}))\, i\mathrm{d}m\mathrm{d}z \right] \mathrm{d}p} = \frac{-\int_{\Phi^{-1}(V)} \triangle H_p(\Phi(\mathbf{r}))\, \mathrm{d}V}{\omega^2 \mu \int_{\Phi^{-1}(V)} H_p(\Phi(\mathbf{r}))\, \mathrm{d}V} \approx \kappa(\Phi(\mathbf{r})) \tag{3.8}$$

with Φ being the described change of basis and $\mathrm{d}V = \mathrm{d}p\mathrm{d}m\mathrm{d}z$ a volume element. The volume integrals in Eq. 3.8 can be transformed back to Cartesian space. The transformation is

$$\mathbf{H}\mathrm{d}p\mathrm{d}m\mathrm{d}z = \mathbf{H} \left| \det \left(\frac{\partial \mathbf{r}}{\partial \tilde{\mathbf{r}}} \right) \right| \mathrm{d}x\mathrm{d}y\mathrm{d}z = \mathbf{H}\, |i|\, \mathrm{d}x\mathrm{d}y\mathrm{d}z = \mathbf{H}\mathrm{d}x\mathrm{d}y\mathrm{d}z. \tag{3.9}$$

Finally, Gauss theorem can be applied to replace $\triangle$

$$\frac{-\int_V \triangle H_p(\mathbf{r})\mathrm{d}V}{\omega^2\mu\int_V H_p(\mathbf{r})\,\mathrm{d}V} = \frac{-\oint_{\partial V} \nabla H_p(\mathbf{r})\mathbf{da}}{\omega^2\mu\int_V H_p(\mathbf{r})\,\mathrm{d}V} \approx \kappa(\mathbf{r}) \tag{3.10}$$

where the integration volume is denoted by V and its surface by ∂V. As pointed out before, the integration in Eq. 3.10 can be performed in Cartesian space. In contrast to the approach of [63], Eq. 3.10 depends only on the measurable H_p, i.e., the positively rotating component of the transmit field, as it is the case in [61, 88, 89, 95]. However, in contrast to the representation shown in [61, 88, 89, 95], Eq. 3.10 does not require the explicit calculation of the second spatial derivative, which is of crucial importance for numerical stability.

In order to obtain separate equations for ε and σ, the differentiation in Eq. 3.10 is executed explicitly, taking into account that $H_p = \mu^{-1}B_p\exp(i\varphi_p)$, where B_p represents the magnitude of the positively rotating component of the RF transmit field (i.e., its "active" component responsible for spin excitation) and φ_p its phase. Thus, Eq. 3.10 now reads

$$\frac{-\oint_{\partial V} \nabla H_p(\mathbf{r})\mathbf{da}}{\omega^2\mu\int_V H_p(\mathbf{r})\,\mathrm{d}V} = \frac{-\oint_{\partial V} e^{i\varphi_p(\mathbf{r})}\nabla B_p(\mathbf{r})\mathbf{da}}{\omega^2\mu\int_V e^{i\varphi_p(\mathbf{r})}B_p(\mathbf{r})\mathrm{d}V} - i\frac{\oint_{\partial V} e^{i\varphi_p(\mathbf{r})}B_p(\mathbf{r})\nabla\varphi_p(\mathbf{r})\,\mathbf{da}}{\omega^2\mu\int_V e^{i\varphi_p(\mathbf{r})}B_p(\mathbf{r})\mathrm{d}V}. \tag{3.11}$$

The following derivations show the details of the calculations leading to Eqs. 3.17 and 3.18. Assuming compartments with constant κ, Eq. 3.11 can be rewritten

$$\kappa(\mathbf{r})\omega^2\mu\int_V B_p(\mathbf{r})e^{i\varphi_p(\mathbf{r})}\mathrm{d}V = -\oint_{\partial V} e^{i\varphi_p(\mathbf{r})}\nabla B_p(\mathbf{r})\mathbf{da} - i\oint_{\partial V} B_p(\mathbf{r})\,e^{i\varphi_p(\mathbf{r})}\nabla\varphi_p(\mathbf{r})\,\mathbf{da}. \tag{3.12}$$

Green's Identity is applied to both parts on the right hand side of Eq. 3.12

$$\oint_{\partial V} e^{i\varphi_p(\mathbf{r})}\nabla B_p(\mathbf{r})\,\mathbf{da} = \int_V e^{i\varphi_p(\mathbf{r})}\triangle B_p(\mathbf{r})\,\mathrm{d}V + i\int_V e^{i\varphi_p(\mathbf{r})}\nabla B_p(\mathbf{r})\cdot\nabla\varphi_p(\mathbf{r})\,\mathrm{d}V \tag{3.13}$$

and

$$
\begin{aligned}
i\oint_{\partial V} & B_p(\mathbf{r}) e^{i\varphi_p(\mathbf{r})} \nabla\varphi_p(\mathbf{r})\,\mathrm{d}\mathbf{a} \\
&= i\int_V B_p(\mathbf{r})\, e^{i\varphi_p(\mathbf{r})} \triangle\varphi_p(\mathbf{r})\,\mathrm{d}V + i\int_V \nabla\Big(B_p(\mathbf{r})\, e^{i\varphi_p(\mathbf{r})}\Big)\cdot\nabla\varphi_p(\mathbf{r})\mathrm{d}V \\
&= i\int_V B_p(\mathbf{r})\, e^{i\varphi_p(\mathbf{r})} \triangle\varphi_p(\mathbf{r})\,\mathrm{d}V + i\int_V e^{i\varphi_p(\mathbf{r})} \nabla B_p(\mathbf{r})\cdot\nabla\varphi_p(\mathbf{r})\,\mathrm{d}V \\
&\quad - \int_V B_p(\mathbf{r})\, e^{i\varphi_p(\mathbf{r})} \left(\nabla\varphi_p(\mathbf{r})\right)^2 \mathrm{d}V.
\end{aligned}
\tag{3.14}
$$

The factor $e^{i\varphi_p(\mathbf{r})}$ can be dropped since the volume integrals in all parts of Eqs. 3.12, 3.13 and 3.14 are identical. Separating real and imaginary parts now yields

$$
\varepsilon(\mathbf{r}) = \left(\omega^2\mu \int_V B_p(\mathbf{r})\,\mathrm{d}V\right)^{-1} \left[\int_V B_p(\mathbf{r})\, e^{i\varphi_p(\mathbf{r})} \left(\nabla\varphi_p(\mathbf{r})\right)^2 \mathrm{d}V - \int_V \triangle B_p(\mathbf{r})\,\mathrm{d}V\right]
\tag{3.15}
$$

and

$$
\sigma(\mathbf{r}) = (\omega\mu V)^{-1} \left(\int_V \triangle\varphi_p(\mathbf{r})\,\mathrm{d}V + 2\int_V \left[\frac{\nabla B_p(\mathbf{r})}{B_p(\mathbf{r})}\cdot\nabla\varphi_p(\mathbf{r})\right]\mathrm{d}V\right).
\tag{3.16}
$$

Applying Gauss theorem to the $\triangle$-terms yields the expressions

$$
\sigma(\mathbf{r}) = (\omega\mu V)^{-1} \left(\oint_{\partial V} \nabla\varphi_p(\mathbf{r})\,\mathrm{d}\mathbf{a} + 2\int_V \left[\nabla \ln\left(B_p(\mathbf{r})\right)\cdot\nabla\varphi_p(\mathbf{r})\right]\mathrm{d}V\right)
\tag{3.17}
$$

$$
\varepsilon(\mathbf{r}) = \left(\omega^2\mu \int_V B_p(\mathbf{r}) dV\right)^{-1} \left[\int_V B_p(\mathbf{r}) \left(\nabla\varphi_p(\mathbf{r})\right)^2 \mathrm{d}V - \oint_{\partial V} \nabla B_p(\mathbf{r})\mathrm{d}\mathbf{a}\right].
\tag{3.18}
$$

Equations 3.17 and 3.18 are stated for compartments with constant κ. Eqs. 3.17 and 3.18 are mathematically equivalent to Eq. 3.10.

An approximate expression for σ and ε can be obtained from Eq. 3.11 by approximating the exponential functions by first order Taylor series and assuming that the variations of B_p along the boundary ∂V are smaller than the variations of $\nabla \varphi_p$. Thereby, the conductivity can be expressed as function of the phase only

$$\sigma(\mathbf{r}) \approx (\omega\mu V)^{-1} \oint_{\partial V} \nabla \varphi_p(\mathbf{r})\, \mathbf{da} \tag{3.19}$$

and the permittivity can be related to the magnitude only

$$\varepsilon(\mathbf{r}) \approx \frac{-\oint_{\partial V} \nabla B_p(\mathbf{r}) \mathbf{da}}{\omega^2 \mu \int_V B_p(\mathbf{r}) dV}. \tag{3.20}$$

Heuristically, Eqs. 3.19 and 3.20 can be motivated by the assumption that equal terms in the fractions of Eq. 3.11 approximately cancel out. I.e., assuming that $\exp(i\varphi_p)$ cancels out yields a quantity, which approximates ε, and assuming that $B_p \exp(i\varphi_p)$ cancels out yields a quantity which approximates σ.

A motivation for Eqs. 3.19 and 3.20 using the concept of plane waves is provided in the next Section 3.3.2.

Equations 3.19 and 3.20 are particularly favorable, since conductivity and permittivity imaging can be split into separate measurements. On the one hand, for conductivity imaging, only the transmit phase has to be determined, skipping the (typically very lenghty) magnitude measurement. On the other hand, in the case of permittivity imaging, only the magnitude has to be determined, skipping an additional phase measurement.

Please note that Eqs. 3.17 and 3.18 as well as Eqs. 3.19 and 3.20 provide absolute values of κ. Any scaling factors of the magnetic field's magnitude cancel out by dividing.

In the following experiments, conductivity and permittivity are determined using the exact Eqs. 3.17 and 3.18, equivalent with Eq. 3.10. Additionally, the applicability of the approximations made in Eqs. 3.19 and 3.20 are investigated.

The same derivations shown here can also be applied to the RF signal emitted not by the RF coil but by the relaxing spins, i.e. to the spatial sensitivity profile H_m of the applied receive coil. Since no exact methods are published to determine receive sensitivities, this study focuses on the use of RF transmit sensitivities.

3.3.2 Motivation of phase/amplitude-based approximations using plane waves

Plane waves in infinite, homogeneous media are used as a simple model for the RF penetration of patients. In this framework, a mathematical motivation can be obtained for the approximations (Eq. 3.19 and 3.20). The plane wave $H_p = \mu^{-1} B_p e^{ikz}$ propagating in z-direction is a solution to Eq. 3.10 with k denoting the wave number. From a comparison of the coefficients, $ik = \pm\sqrt{\omega^2\mu\varepsilon - i\omega\mu\sigma}$ is obtained. The root can be written as a sum of real and imaginary part (w.l.o.g. only one branch is used)

$$ik = \sqrt{\frac{\omega\mu}{2}}\left[-\sqrt{\omega|\kappa| + \omega\varepsilon} + i\sqrt{\omega|\kappa| - \omega\varepsilon}\right]. \tag{3.21}$$

The imaginary part of Eq. 3.21 describes the transmit phase φ_p and the real part describes the influence on the magnitude B_p. Two regimes of applicability of the approximations Eqs. 3.19 and 3.20 will be investigated.

I) First, the observed applicability of the approximations Eqs. 3.19 and 3.20 is motivated for the regime $\sigma \approx \omega\varepsilon$, valid for most human tissue types as well as for cylinder A used in the presented phantom study (Fig. 3.3). To this goal, the impact of σ and ε on the amplitude B_p is estimated by the corresponding derivatives of the real part of Eq. 3.21

$$\frac{\partial\Re(ik)}{\omega\partial\varepsilon} = -\sqrt{\frac{\omega\mu}{8}}\frac{1}{\sqrt{\omega|\kappa| + \omega\varepsilon}}\left(\frac{\omega\varepsilon}{\omega|\kappa|} + 1\right), \tag{3.22}$$

$$\frac{\partial\Re(ik)}{\partial\sigma} = -\sqrt{\frac{\omega\mu}{8}}\frac{\sigma}{\omega|\kappa|\sqrt{\omega|\kappa| + \omega\varepsilon}}. \tag{3.23}$$

Using $\sigma = \varepsilon\omega$ and forming the ratio of the norms of Eqs. 3.22 and 3.23 yields

$$\frac{\left|\frac{\partial\Re(ik)}{\omega\partial\varepsilon}\right|_{\sigma=\omega\varepsilon}}{\left|\frac{\partial\Re(ik)}{\partial\sigma}\right|_{\sigma=\omega\varepsilon}} = 1 + \sqrt{2} > 1. \tag{3.24}$$

Thus, changes in the real part and hence in the magnitude B_p are predominately induced by ε. A similar derivation can be applied to the imaginary part of Eq. 3.21. The derivatives are

$$\frac{\partial \Im(ik)}{\partial \sigma} = \sqrt{\frac{\omega\mu}{8}} \frac{\sigma}{\omega|\kappa|\sqrt{\omega|\kappa| - \omega\varepsilon}} \tag{3.25}$$

$$\frac{\partial \Im(ik)}{\omega\partial \varepsilon} = \sqrt{\frac{\omega\mu}{8}} \frac{1}{\sqrt{\omega|\kappa| - \omega\varepsilon}} \left(\frac{\omega\varepsilon}{\omega|\kappa|} - 1\right). \tag{3.26}$$

The ratio of Eqs. 3.25 and 3.26 for $\sigma = \varepsilon\omega$ can be computed, yielding

$$\frac{\left|\frac{\partial \Im(ik)}{\partial \sigma}\right|_{\sigma=\omega\varepsilon}}{\left|\frac{\partial \Im(ik)}{\omega\partial \varepsilon}\right|_{\sigma=\omega\varepsilon}} = \frac{1}{\left|1-\sqrt{2}\right|} > 1. \tag{3.27}$$

Thus, changes in the imaginary part and hence in the transmit phase φ_p are predominately induced by σ.

II) Second, the increased accuracy of the approximations Eqs. 3.19 and 3.20 for the regimes $\sigma \gg \omega\varepsilon$ (body fluids / cylinder B) and $\omega\varepsilon \gg \sigma$ will be motivated, again using the model of plane waves. For $\omega\varepsilon \gg \sigma$, the approximation $\omega|\kappa| = \sqrt{\omega^2\varepsilon^2 + \sigma^2} \approx \omega\varepsilon$ can be used in Eq. 3.21, which yields

$$\Re(ik) \approx -\omega\sqrt{\mu\varepsilon}. \tag{3.28}$$

This equation motivates amplitude based permittivity imaging for $\omega\varepsilon \gg \sigma$, since its real part (and thus, the amplitude of H_p) is influenced by permittivity only. For $\sigma \gg \omega\varepsilon$, the approximation $\omega|\kappa| - \omega\varepsilon \approx \sigma$ is used to yield

$$\Im(ik) \approx \sqrt{\frac{\omega\mu\sigma}{2}}. \tag{3.29}$$

This equation motivates phase-based conductivity imaging for $\sigma \gg \omega\varepsilon$, since its imaginary part (and thus, the phase of H_p) is influenced by the conductivity only.

3.3.3 EPT reconstruction and dielectric boundaries

In the derivation of the central equations, constant dielectric properties within the reconstruction volume have been assumed. In principle, the theory of conductivity and permittivity imaging can be expanded to include non-constant κ. Instead of cancelling the electric fields inside Eq.

3.3, Stokes theorem and the vector identity $\nabla \times \nabla \times \mathbf{H} = \nabla(\nabla \cdot \mathbf{H}) - \triangle \mathbf{H} = -\triangle \mathbf{H}$ can be used to obtain

$$\frac{-\int_A \triangle \mathbf{H}(\mathbf{r})\,\mathbf{da}}{\omega^2 \mu \int_A \mathbf{H}(\mathbf{r})\mathbf{da}} = \frac{\int_A (\nabla \times \kappa(\mathbf{r})\mathbf{E}(\mathbf{r}))\,\mathbf{da}}{\omega^2 \mu \int_A \mathbf{H}(\mathbf{r})\mathbf{da}} \tag{3.30}$$

instead of Eq. 3.4. Evaluating the argument of the nominator on the right side of Eq. 3.30 yields

$$\begin{aligned}
(\nabla \times \kappa(\mathbf{r})\mathbf{E}(\mathbf{r})) &= [\kappa(\mathbf{r})(\nabla \times \mathbf{E}(\mathbf{r})) + (\nabla\kappa(\mathbf{r}) \times \mathbf{E}(\mathbf{r}))] \\
&= \kappa(\mathbf{r})\mathbf{H}(\mathbf{r}) + \left(\frac{\nabla\kappa(\mathbf{r}) \times (\nabla \times \mathbf{H}(\mathbf{r}))}{\omega^2\mu\kappa(\mathbf{r})}\right) \\
&= \kappa(\mathbf{r})\mathbf{H}(\mathbf{r}) + \frac{(\nabla\mathbf{H}(\mathbf{r}))\cdot\nabla\kappa(\mathbf{r}) - \nabla\kappa(\mathbf{r})(\nabla\mathbf{H}(\mathbf{r}))}{\omega^2\mu\kappa(\mathbf{r})}
\end{aligned} \tag{3.31}$$

where $\nabla\mathbf{H}(\mathbf{r})$ is the Jacobian of the magnetizing field vector $\mathbf{H}$. Ampere's law $\nabla \times \mathbf{H}(\mathbf{r}) = \omega^2\mu\kappa(\mathbf{r})\mathbf{E}(\mathbf{r})$ has been used in the second step and in the third step the vector identity $\mathbf{C} \times (\nabla \times \mathbf{D}) = \nabla_D(\mathbf{C}\cdot\mathbf{D}) - (\mathbf{C}\cdot\nabla)\mathbf{D} = (\nabla\mathbf{D})\cdot\mathbf{C} - \mathbf{C}\cdot(\nabla\mathbf{D})$, where the subscript notation ∇_D means that the subscripted gradient operates on only the vector $\mathbf{D}$. Since both integrals on the right side of Eq. 3.30 are the same, the derivation by $\mathbf{H}$ can be executed

$$\frac{-\int_A \triangle \mathbf{H}(\mathbf{r})\,\mathbf{da}}{\omega^2 \mu \int_A \mathbf{H}(\mathbf{r})\mathbf{da}} = \kappa(\mathbf{r}) + \int_A \left(\frac{\nabla\kappa(\mathbf{r}) \times (\nabla \times \mathbf{H}(\mathbf{r}))}{\mathbf{H}(\mathbf{r})\omega^2\mu\kappa(\mathbf{r})}\right)\mathbf{da}. \tag{3.32}$$

Using the same coordinate transformation Φ and additional integration as described in Section 3.3.1 can be used to obtain

$$\frac{-\int_{\partial V} \nabla H_p(\mathbf{r})\,\mathbf{da}}{\omega^2 \mu \int_V H_p(\mathbf{r})\mathrm{d}V} = \kappa(\mathbf{r}) + \int_V \frac{(\nabla\kappa(\mathbf{r}) \times (\nabla \times \mathbf{H}(\mathbf{r})))_p}{H_p(\mathbf{r})\omega^2\mu\kappa(\mathbf{r})}\mathrm{d}V \tag{3.33}$$

where the approximate version Eq. 3.10 has been obtained plus an additional term that depends on κ and $\mathbf{H}$ and their derivatives. The subscript p denotes the positively rotating component. Eq. 3.33 is correct also for variations of κ. On the other hand, all components of the magnetic field are required in the correction term of the exact form. In this work, only the approximate Eq. 3.10 will be used. A systematic investigation of the influence of neglecting the unmeasurable components in the correction term could be combined with the development of an iterative reconstruction algorithm to solve the differential Eq. 3.33.

3.3.4 Magnetic field magnitude and phase imaging

EPT applies the equations discussed above to the RF field generated by a transmit (Tx) / receive (Rx) RF coil of a standard MR system. Its magnitude B_p can be measured via corresponding mapping techniques (see, e.g., Chapter 2 of this work). However, as discussed in the last paragraph, also the knowledge of the transmit phase φ_p is required for EPT. As to the authors' knowledge, there is no method published to exactly determine φ_p. Assuming a repetition time (TR) and echo time (TE) of TR $\gg T_1 \gg$ TE, the signal obtained by an RF coil in MRI is given by [103],

$$S(\alpha,\mathbf{r}) = V_1 M_0(\mathbf{r})\,|B_m^{'}(\mathbf{r})\,|\exp(i\varphi(\mathbf{r}))\sin(V_2\alpha|B_p(\mathbf{r})\,|) \qquad (3.34)$$

with nominal flip angle α and V_1 , V_2 system-dependent constants. Relaxation effects and spin density are taken into account in the factor M_0. Here, $B_m^{'}$ is the magnitude of the receive sensitivity of the RF coil. The prime indicates that the coil used for Rx can be different from the Tx coil, or, if the Tx coil is also used for Rx, it might not be identically driven (e.g. with switched polarization).

The signal phase $\varphi = \varphi_p + \varphi_m^{'}$ contains both, the Tx φ_p and the Rx phase $\varphi_m^{'}$. Assuming applicability of Eq. 3.19, any superposition of two physical phases leads to exactly twice the reconstructed conductivity due to the linearity of the equation. Using the exact formula for conductivity, the following expression is obtained

$$\begin{aligned}\widetilde{\sigma}(\mathbf{r}) &= (\omega\mu V)^{-1}\left(\oint_{\partial V}\nabla\left(\varphi_p(\mathbf{r})+\varphi_m^{'}(\mathbf{r})\right)(\mathbf{r})\,\mathrm{d}\mathbf{a}\right.\\ &\left.+2\int_V\left[\nabla\ln\left(B_p(\mathbf{r})\right)\cdot\nabla\left(\varphi_p(\mathbf{r})+\varphi_m^{'}(\mathbf{r})\right)\right]\mathrm{d}V\right)\\ &= \sigma(\mathbf{r})+(\omega\mu V)^{-1}\left(\oint_{\partial V}\nabla\varphi_m^{'}(\mathbf{r})(\mathbf{r})\,\mathrm{d}\mathbf{a}\right.\\ &\left.+2\int_V\left[\nabla\ln\left(B_p(\mathbf{r})\right)\cdot\nabla\varphi_m^{'}(\mathbf{r})\right]\mathrm{d}V\right)\\ &\approx 2\sigma(\mathbf{r}). \end{aligned} \tag{3.35}$$

The result is only approximately twice the conductivity because of the mixing term involving B_p instead of $B_m^{'}$. This justifies the introduction of a phase factor of 0.5 to be applied to the reconstructed conductivity for arbitrary combinations of Tx and Rx (surface) RF coils in the scope of Eq. 3.19. In cases where the phase-based approximation and its linearity is not fulfilled sufficiently, the Rx phase might be removed with the help of a concept called CLEAR (see Section 4.2.1 in the next chapter). For permittivity imaging the impact is expected to be neglegible, since phase does not enter Eq. 3.20. Using the exact expression yields

$$\begin{aligned}\widetilde{\varepsilon}(\mathbf{r}) &= \left(\omega^2\mu_0\int_V B_p(\mathbf{r})dV\right)^{-1}\left[\int_V B_p(\mathbf{r})\left(\nabla\left(\varphi_p(\mathbf{r})+\varphi_m^{'}(\mathbf{r})\right)\right)^2\mathrm{d}V\right.\\ &\left.-\oint_{\partial V}\nabla B_p(\mathbf{r})\mathrm{d}\mathbf{a}\right]\\ &= \varepsilon(\mathbf{r})+\left(\omega^2\mu_0\int_V B_p(\mathbf{r})dV\right)^{-1}\left[\int_V B_p(\mathbf{r})\left(\nabla\varphi_m^{'}(\mathbf{r})\right)^2\mathrm{d}V\right.\\ &\left.+2\int_V B_p(\mathbf{r})\nabla\varphi_m^{'}(\mathbf{r})\cdot\nabla\varphi_p(\mathbf{r})\,\mathrm{d}V\right]\\ &\approx \varepsilon(\mathbf{r}) \end{aligned} \tag{3.36}$$

where it has been assumed that the phase related parts in permittivity imaging are neglegible.

An attempt to directly measure Tx phase is made in Section 4.2.1 and tested in phantom experiments. It is based on the approximation that $\varphi_p \approx 0.5\varphi$. This relation is exactly valid in an empty volume coil driven in

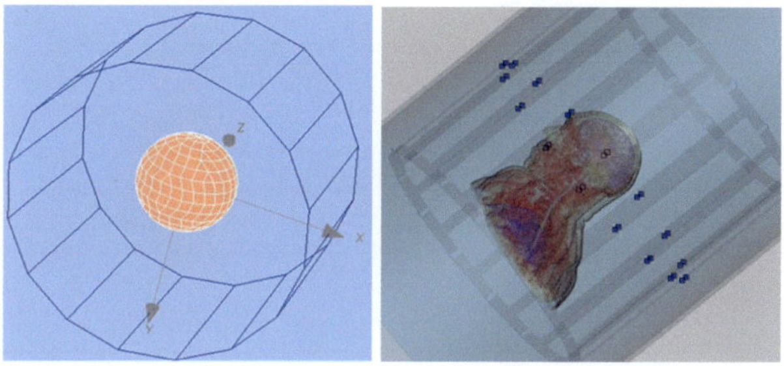

(a) **(b)**

Figure 3.2: Sketch of objects used for the simulations presented in this work. (a) Spherical phantom with homogeneous electric properties. (b) Realistic head model of the visible human for FDTD simulations.

quadrature mode, and turned out to be a good approximation for loaded coils at 64 MHz ([63], [95]) as well as higher field strengths [104]. The use of spin echo (SE) sequences in phase imaging assures that unwanted phase contributions from, e.g., main field strength (B_0) inhomogeneities are removed by the refocusing pulse.

3.4 Methods of EPT validations

3.4.1 Simulations

Two simulation studies were conducted to assess Eqs. 3.17 and 3.18, and particularly, the validity of the approximation given in Eqs. 3.19 and 3.20. Firstly, a spherical phantom (diameter = 20 cm) with uniform conductivity and permittivity was placed at the isocenter of a body coil (Fig. 3.2).

The coil was driven in quadrature mode at 64 MHz, and the resulting B_p magnitude and phase φ_p were computed using the software package CONCEPT II (Technical University Hamburg–Harburg, Department of Theoretical Electrical Engineering, Germany). CONCEPT II is a numerical simulation program of electromagnetic parameters on metallic structures and homogeneous media. It is based upon electric field integral equations and the method of moments [105]. For a given set of electric properties, simulated B_p and φ_p were used for the reconstruction of permittivity and conductivity according to the exact expressions Eqs. 3.17 and 3.18 and

	Conductivity [S/m]		Permittivity [ε_0]		Input value [97]	
ROI	Exact formula	Phase-based	Exact formula	Magnitude-based	σ [S/m]	ε [ε_0]
WM	0.30 ± 0.03	0.33 ± 0.04	68.4 ± 3.8	62.3 ± 5.8	0.29	67.8
GM	0.52 ± 0.06	0.57 ± 0.07	100 ± 8.9	89.1 ± 10.6	0.51	97.4
CSF	2.09 ± 0.12	2.19 ± 0.08	103.8 ± 11.8	n/a	2.07	97.3

Table 3.2: Analysis of FDTD results. Quantitative comparison of average values of conductivity and permittivity inside different regions of interest. Quantities were reconstructed from FDTD data and the exact reconstruction results compared with the apporximations discussed in this chapter. Corresponding images are shown in Figs. 3.5 and 3.6.

the approximations Eqs. 3.19 and 3.20. The simulation was repeated for $0\,\text{S/m} < \sigma < 2.5\,\text{S/m}$ and $0\,\text{S/m} < \omega\varepsilon < 1.2\,\text{S/m}$ in steps of $0.1\,\text{S/m}$, covering the whole range of electric properties of human tissue at a main field strength (B_0) of 1.5 T. The (NRMSE) of the resulting σ and ε was calculated for each set of electrical properties.

Secondly, an FDTD study was performed using the visible human model in Semcad (Schmid & Partner Engineering AG, Zürich, Switzerland). The head of the visible human was placed at the isocenter of a body coil at 1.5 T (Fig. 3.2). The coil was driven in a resonant mode tuned to 64 MHz. The electric properties of the head were set according to [97] and are shown in Table 3.2.

A resolution of $1 \times 1\,\text{mm}^2$ in-plane was used to simulate the magnetic fields at 64 MHz. Again, magnitude B_p and phase φ_p of the resulting transmit field were used for exact and approximate reconstruction according to Eqs. 3.17, 3.18 and Eqs. 3.19, 3.20, respectively. The NRMSE was computed for 15 coronal slices to compare exact and approximate reconstruction results. A measure of the quality of the approximation within a slice is provided via the ratio of retained and neglected components

$$Q_\sigma = \log\left(\frac{\oint_{\partial V} \nabla\varphi_p(\mathbf{r})\,\mathbf{da}}{2\int_V \left[\nabla \ln\left(B_1(\mathbf{r})\right)\cdot\nabla\varphi_p(\mathbf{r})\right]\mathrm{d}V}\right) \tag{3.37}$$

$$Q_\varepsilon = \log\left(\frac{\oint_{\partial V} \nabla B_1(\mathbf{r})\mathrm{d}\mathbf{a}}{\int_V B_1(\mathbf{r})\left(\nabla\varphi_p(\mathbf{r})\right)^2 \mathrm{d}V}\right). \tag{3.38}$$

From the FDTD data, different regions of interest (ROI) with constant conductivity and permittivity (i.e., white matter (WM) / grey matter (GM) / cerebrospinal fluid (CSF)) were segmented and reconstructed separately. For computation of the derivatives in Eqs. 3.17, 3.18, 3.19, and 3.20, Savitzky-Golay filtering [106] was used, i.e., a set of convolution coefficients, obtained by a local polynomial regression, was applied to the simulated magnitude and/or phase maps. In this study, the convolution coefficients were based on a parabola fitted to $I = 2$ voxels on each side of the targeted voxel, both in in-plane and in through-plane direction. For computation of the integrals, $J = 1$ adjacent voxel was used in each direction. A fixed size of the described convolution matrix would frequently lead to undefined pixels along ROI boundaries. Therefore, the matrix size was adjusted dynamically to $I = 1$ in the vicinity of ROI boundaries to minimize the number of undefined voxels. Still occurring gaps in the conductivity and permittivity maps were closed by a region growing algorithm.

3.4.2 Phantom experiments

A phantom study was performed to demonstrate the feasibility and accuracy of the proposed method. Two cylinders (diameter 7.5 cm and height 6 cm) filled with saline solution with known conductivity and permittivity were imaged using a standard 1.5 T scanner (Philips Healthcare, Best, The Netherlands). Electric conductivity values were determined with a 4-ring potentiometric probe (HI8733, Hanna Instruments, USA). Permittivity was determined using different mixing ratios of water and 2-propanol. Both cylinders with dielectric properties ($\sigma_A = 0.42\,\mathrm{S/m}$, $\varepsilon_A = 50$) and ($\sigma_B = 1.35\,\mathrm{S/m}$, $\varepsilon_B = 75$) were placed in a Tx/Rx head coil with a feet-head offset of 1 cm. T_1 of the phantoms was reduced to approximately 300 ms using 2.5 ml Magnevist (Bayer Schering Pharma AG, Berlin, Germany) per liter of phantom fluid. For B_p mapping, the multiple TR B_p/T_1 mapping (MTM) sequence was used (see Chapter 2). MTM TR sets were $\mathrm{TR}_{11,12} = 30/80\,\mathrm{ms}$, $\mathrm{TR}_{21,22} = 30/130\,\mathrm{ms}$, $\mathrm{TR}_{31,32} = 30/185\,\mathrm{ms}$, and $\mathrm{TR}_{41,42} = 30/200\,\mathrm{ms}$ with nominal flip angle 70° and total acquisition time of 23 min. For φ_p mapping, the phase of a 90° spin echo (SE) image acquired in 5 : 45 min was cut in half to separate transmit from receive phase as discussed above. The voxel size was set to $1.5 \times 1.5 \times 5\,\text{mm}^3$. Conductivity reconstruction was performed according to Eq. 3.17 and using the approximate relation Eq. 3.19, applying a kernel of $I = 5$ voxels for in-plane derivatives and $I = 4$ voxels for through-plane derivatives. Permittivity reconstruction was performed according to Eq. 3.18 and using the approximate relation Eq. 3.20 using $I = 8$ voxels in-plane and $I = 6$ voxels through-plane. For the integral, $J = 1$ voxel was used in both cases. For each of the 8 slices, reconstructed conductivity and permittivity were averaged inside a circular region of interest and compared with the expected values.

In another study, the dynamic range of EPT has been addressed. The dielectric properties of one of the cylinders were varied covering the whole physiologic range. Different salt concentrations were used yielding conductivity values $\sigma = (0.01/0.42/0.55/0.85/1.35/2.05)\,\text{S/m}$ at permittivity $\varepsilon = 80$. At conductivity $\sigma \approx 0\,\text{S/m}$ different permittivities were measured $\varepsilon = (18/26/34/42/49/57/65/72/80)\,\varepsilon_0$. Please note, that using water propanol mixtures, no permittivity value higher than water can be made. For imaging, the same sequences were applied and the reconstruction was performed accordingly. For permittivity imaging, the sequences were applied at $B_0 = 3\,\text{T}$.

3.5 *In vivo* application of EPT

3.5.1 Volunteer experiments

In vivo experiments were conducted with healthy volunteers using a standard 1.5 T scanner (Philips Healthcare, Best, The Netherlands). Written consent was obtained from all volunteers. The heads of the volunteers were imaged using a transmit/receive head coil. As elaborated in the Section “Discussion” (3.7), permittivity maps tend to show lower SNR than corresponding conductivity maps. To compensate for this effect in the comparison of Eqs. 3.17 and 3.18 with Eqs. 3.19 and 3.20, two separate experiments for *in vivo* permittivity and conductivity imaging were performed as described below.

1. Conductivity imaging experiments

For both magnitude B_p and phase φ_p mapping, the voxel size was set to

$1.6 \times 1.6 \times 2\,\text{mm}^3$. Eleven coronal slices were acquired in 3D acquisition mode. For phase mapping, a spin echo (SE) sequence was used with TR = 200 ms, TE = 8 ms, and nominal flip angle 90°, resulting in a total imaging time of 5 min. To estimate the transmit phase φ_p, the phase of the SE image was cut in half to separate transmit from receive phase as discussed above. For B_p magnitude mapping, the MTM sequence (see Chapter 2) was used. For the current study, two different pairs of TR were performed ($TR_{11,12} = 15/215\,\text{ms}$, $TR_{21,22} = 15/315\,\text{ms}$), leading to a total scan time of 11 min. Just as for the FDTD study, three ROIs were defined based on the anatomy prior to reconstruction. Simple threshold segmentation has been used to distinguish between WM, GM, and CSF. Average values and standard deviation inside these compartments were calculated from the reconstructed electric conductivity. Conductivity reconstruction was performed according to the exact formula Eq. 3.17 as well as the phase-based approximation Eq. 3.19. Reconstruction parameters were $I = 6$ pixels on each side of the targeted pixel for in-plane derivatives and $I = 5$ pixel for derivatives in through-plane direction. Two adjacent lines parallel to the targeted line were averaged. The size of the volume integral was set to $J = 1$ pixel in each direction. Values for I and J are determined by visual inspection of the reconstruction results and represent a tradeoff between SNR and resolution of the reconstructed maps.

2. Permittivity imaging experiments

As for the conductivity imaging, B_p was measured via MTM (Chapter 2) and φ_p via an SE sequence. However, to compensate for the relatively low SNR usually found for B_p maps, the voxel size was increased to $3.1 \times 3.1 \times 15\,\text{mm}^3$. Moreover, three instead of two TR pairs were used for MTM ($TR_{11}/TR_{12} = 20/363\,\text{ms}$, $TR_{21}/TR_{22} = 20/257\,\text{ms}$, $TR_{31}/TR_{32} = 30/67\,\text{ms}$), leading to a total scan time of 13 min for 16 slices. The SE sequence took 4 min using TR = 200 ms at a nominal flip angle of 90°. Permittivity reconstruction was performed according to the exact formula Eq. 3.18 and using the magnitude-based approximation Eq. 3.20 using the same reconstruction parameters as for the conductivity reconstruction (see last paragraph).

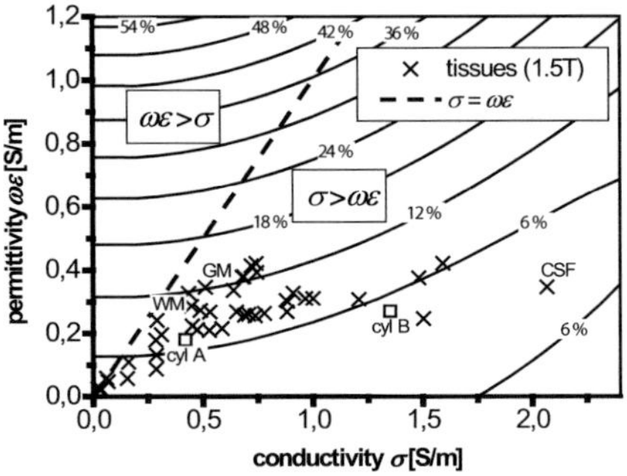

Figure 3.3: Conductivity NRMSE introduced by phase-based approximation for the simulation setup shown in Fig. 3.2a. The dashed line separates the two regions $\omega\varepsilon > \sigma$ and $\sigma > \omega\varepsilon$. Highest errors in reconstructed conductivity are found for $\omega\varepsilon \gg \sigma$. Dielectric properties of human tissue are indicated by crosses. For most tissues, NRMSE is of the order of 10%.

3.6 Results

3.6.1 Simulations

For a homogeneous, spherical phantom, the reconstruction of electric properties according to the exact formulas Eqs. 3.17 and 3.18 did not show any significant errors (NRMSE $\ll 1\%$). The errors introduced applying the approximate Eqs. 3.19 and 3.20 to a spherical phantom are shown in Figs. 3.3 and 3.4.

The approximation for conductivity seems to be more accurate the more the relation $\sigma \gg \omega\varepsilon$ is fulfilled. The approximation for permittivity seems to be more accurate the more the relation $\omega\varepsilon \gg \sigma$ is fulfilled For physiological tissue values at 1.5 T (indicated in the plots by crosses), roughly $2\omega\varepsilon < \sigma < 10\omega\varepsilon$ can be found. Consequently, the error in estimating the electric conductivity from phase information only is of the order of $5-15\,\%$. The error in estimating the permittivity from magnitude information only is of the order of $10-20\,\%$ for most tissues, however, increases significantly for highly conductive tissues, e.g. the body fluids CSF or blood. The comparison of full and approximated reconstruction using the visible human model shows little qualitative deviation (Fig. 3.5

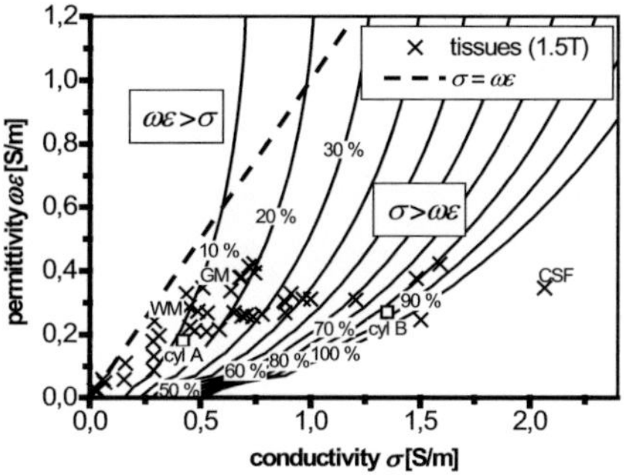

Figure 3.4: Permittivity NRMSE introduced by magnitude-based approximation for the simulation setup shown in Fig. 3.2a. The dashed line separates the two regions $\omega\varepsilon > \sigma$ and $\sigma > \omega\varepsilon$. Highest errors in reconstructed permittivity are found for $\sigma \gg \omega\varepsilon$. Dielectric properties of human tissue are indicated by crosses. For most tissues, except for highly conductive body fluids like blood and CSF, NRMSE is of the order of 20%.

and 3.6).

Residual errors in the exact reconstruction case are due to numerical effects. Fitting a 2nd order polynomial is particularly suitable for constant conductivity (permittivity) since in this case solving the inverse of Eq. 3.19 (Eq. 3.20) yields quadratic phase φ_p (magnitude B_p). This explains the high accuracy achieved in simulations of the homogeneous sphere and lower accuracy achieved in simulations of the heterogeneous head model. In the latter case, the use of higher order polynomials improves the results. However, higher order polynomials suffer from higher noise sensitivity, and thus, only 2nd order polynomials are used throughout this study. The quantitative NRMSE introduced by the approximations is of the order of 10 % for conductivity and 15 % for permittivity reconstructions (Fig. 3.7). In the case of permittivity imaging, the NRMSE is related to the CSF content present in the reconstructed slice, i.e., the more CSF present, the higher the error obtained, as expected from Fig. 3.4. Quantitative values of conductivity and permittivity for full and approximate reconstruction are shown in Table 3.2. In the case of magnitude-based permittivity reconstruction, no physically meaningful value for CSF could be obtained.

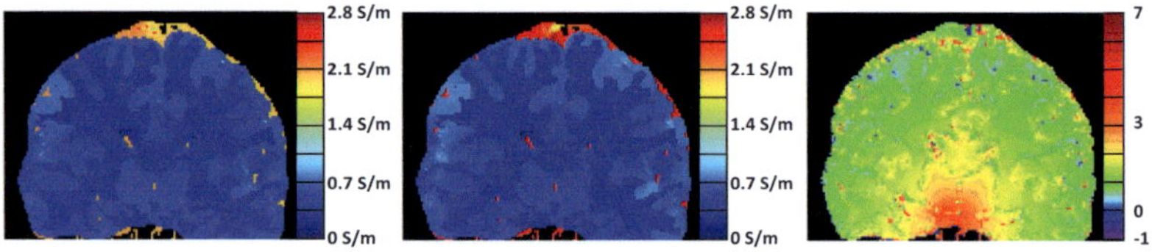

(a) (b) (c)

Figure 3.5: Conductivity reconstruction of FDTD simulation data for the setup shown in Fig. 3.2b. (a) Exact reconstruction using Eq. 3.17, i.e., based on complex H_p. (b) Approximated reconstruction using Eq. 3.19, i.e., based only on the transmit phase φ_p. (c) Relative accuracy according to Eq. 3.37 of the approximation underlying Fig. 3.5b. A corresponding, quantitative analysis can be found in Table 3.2.

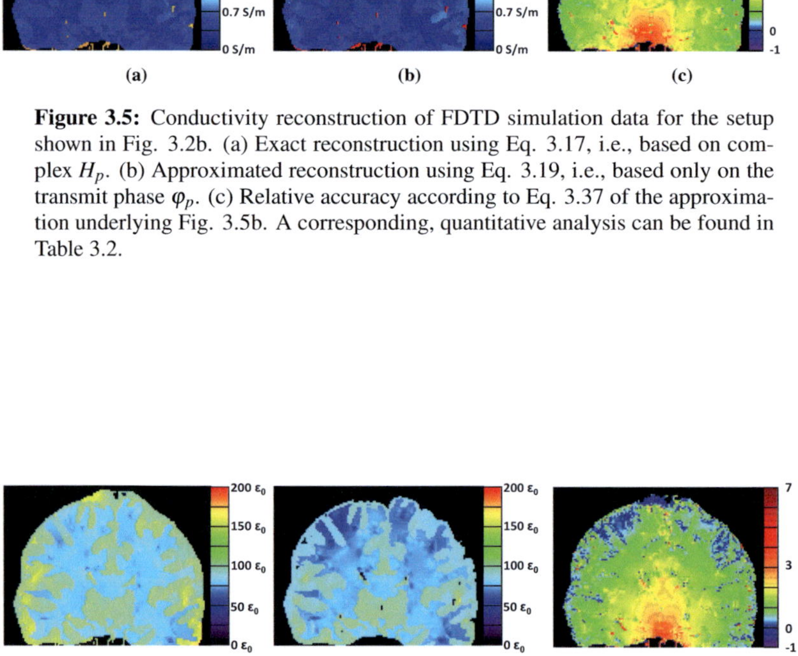

(a) (b) (c)

Figure 3.6: Permittivity reconstruction of FDTD simulation data for the setup shown in Fig. 3.2b. (a) Exact reconstruction using Eq. 3.18, i.e., based on complex H_p. (b) Approximated reconstruction using Eq. 3.20, i.e., based only on B_p magnitude. CSF is not shown in (b) since magnitude-based reconstruction of CSF permittivity violates $\omega\varepsilon \gg \sigma$, and thus, does not yield a meaningful value (cf. Fig. 3.4). (c) Relative accuracy according to Eq. 3.38 of the approximation underlying Fig. 3.6b. A corresponding, quantitative analysis can be found in Table 3.2.

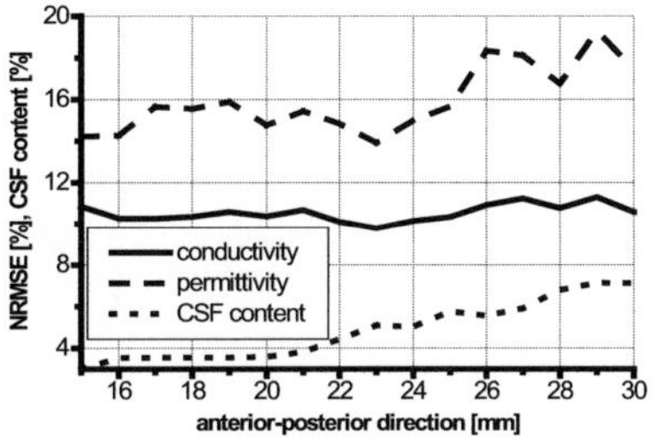

Figure 3.7: Comparison between full and approximated reconstruction of simulated brain data. The NRMSE introduced via magnitude-based permittivity reconstruction are closely related to CSF content in corresponding slices. Increasing CSF content indicates less accurate magnitude-based permittivity reconstructions.

3.6.2 Phantom experiments

Quantitative results of the phantom experiments are shown in Figs. 3.8 and 3.9.

Good agreement is observed between reconstructed and expected values for conductivity and permittivity over the whole 3D volume (Fig. 3.8). Errors introduced by the approximations made in Eqs. 3.19 and 3.20 were not significant except for magnitude-based permittivity imaging of cylinder B. This can be explained from the fact that for cylinder B $\sigma_B/\omega\varepsilon_B \approx 1.35/0.27 = 5$ is found unsuitable for permittivity approximation (Fig. 3.4), whereas $\sigma_A/\omega\varepsilon_A \approx 0.42/0.18 = 5$ for cylinder A allows both, phase-based (Fig. 3.3) as well as amplitude-based reconstruction (Fig. 3.4). Good agreement between exact EPT reconstruction and independently measured values could also be confirmed over a wide range of different values for conductivity and permittivity (Fig. 3.9). Errors introduced by the approximations made in Eqs. 3.19 and 3.20 were not significant within the tested range of dielectric properties.

3.6.3 Volunteer experiments

In vivo conductivity and permittivity images are shown in Fig. 3.10 and Fig. 3.11. In both cases, only marginal differences between the differently reconstructed images are observed. A quantitative analysis of three ROIs is shown in Table 3.3. Average values inside ROIs obtained for permit-

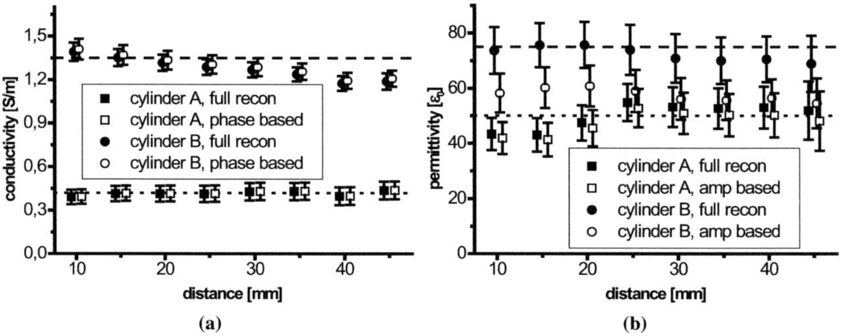

Figure 3.8: Phantom experiments. (a) Conductivity imaging according to the presented method yields accurate results. Phase-based conductivity reconstruction results in negligible deviation from the full reconstruction result. (b) Permittivity imaging using the presented method yields accurate results, however superimposed by stronger noise. Amplitude based permittivity reconstruction works well for cylinder A and fails for cylinder B as expected due to $\sigma_B/\omega\varepsilon_B = 1.35/0.27 = 5$. Data points for identical slices have been shifted slightly to improve readability.

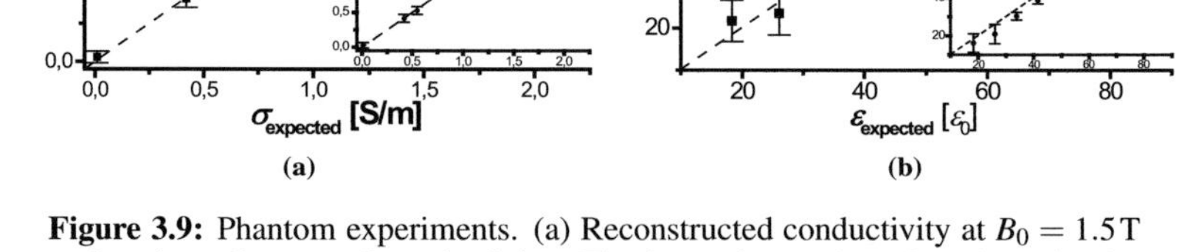

Figure 3.9: Phantom experiments. (a) Reconstructed conductivity at $B_0 = 1.5\,\mathrm{T}$ as function of expected conductivity. The inset shows phase-based conductivity reconstruction. (b) Reconstructed permittivity at $B_0 = 3\,\mathrm{T}$ as function of expected permittivity. Permittivity values, although obtained at higher field strength, yielded lower SNR. The inset shows magnitude-based permittivity reconstruction.

	Conductivity [S/m]		Permittivity [ε_0]	
ROI	Exact formula	Phase-based approximation	Exact formula	Magnitude-based approximation
WM	0.39 ± 0.15	0.43 ± 0.15	72 ± 64	63 ± 66
GM	0.69 ± 0.14	0.72 ± 0.15	103 ± 69	91 ± 70
CSF	1.75 ± 0.34	1.82 ± 0.37	104 ± 21	98 ± 20

Table 3.3: Analysis of *in vivo* study. Quantitative comparison of average values of conductivity and permittivity inside different regions of interest. Values were reconstructed from measured data of a healthy volunteer. Corresponding images are shown in Figs. 3.10 and 3.11.

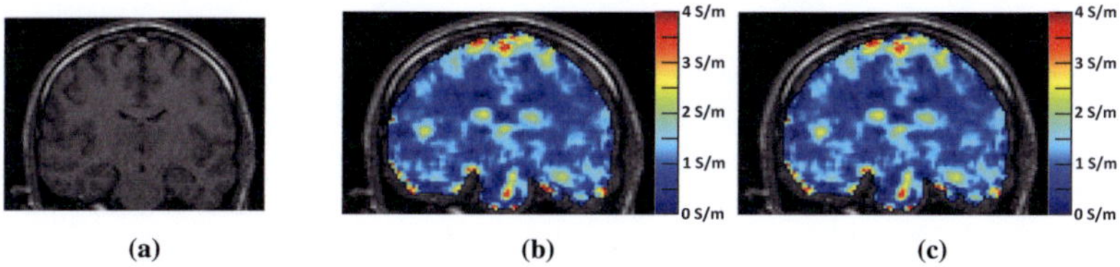

Figure 3.10: Brain conductivity study. (a) SE image acquired to determine φ_p. (b) Exact reconstruction using Eq. 3.17, i.e., based on complex H_p. (c) Approximated reconstruction using Eq. 3.19, i.e., based only on the transmit phase φ_p. The correlation of (a-c) is analyzed quantitatively in Table 3.3.

tivity and conductivity are in agreement with literature values [97]. As already observed in the simulation study, the phase-based conductivity reconstruction leads to systematically higher values than in the exact reconstruction case and the magnitude-based permittivity reconstruction leads to lower values compared with the exact reconstruction. The considerable errors found for permittivity imaging arise from the large scale calculus operations smearing over different compartments.

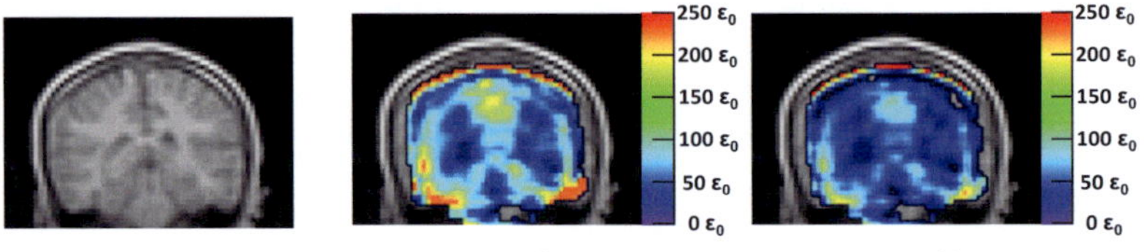

Figure 3.11: Brain permittivity study. (a) FFE image acquired to determine B_p. (b) Exact reconstruction using Eq. 3.18, i.e., based on complex H_p. (c) Approximated reconstruction using Eq. 3.20, i.e., based only on B_p magnitude. The correlation of (a-c) is analyzed quantitatively in Table 3.3.

3.6.4 High-field application of EPT

SNR problems as encountered with permittivity imaging at 1.5 T can be circumvented by using a higher field strength B_0. At $B_0 = 3\,\mathrm{T}$, two experiments with healthy volunteers have been conducted: A first experiment designed to test the impact of higher field strength on permittivity imaging and a second experiment to apply phase-based conductivity imaging with high resolution.

For permittivity imaging, an MTM sequence as described in the previous section has been used with repetition times $TR_{11}/TR_{12} = 30/230\,\mathrm{ms}$, $TR_{21}/TR_{22} = 30/150\,\mathrm{ms}$, $TR_{31}/TR_{32} = 30/90\,\mathrm{ms}$ leading to imaging times $5:57\,\mathrm{min}$, $4:07\,\mathrm{min}$ and $2:40\,\mathrm{min}$, respectively. For phase mapping, a turbo SE sequence with turbo factor 2, $TR = 460\,\mathrm{ms}$, and acquisition time $6:47\,\mathrm{min}$ was used. With both sequences, 22 slices were acquired with $2.5 \times 2.5\,\mathrm{mm}^2$ in-plane resolution and 10 mm slice thickness. Permittivity reconstruction was performed using 4 pixels for the in-plane derivatives and 2 pixels in through-plane direction. Permittivity imaging results are shown in Fig. 3.12b. Comparing to permittivity imaging at 1.5 T, the map reveals more anatomic details. The resolution is much higher, maintaining SNR. Hence, higher field is particularly beneficial for permittivity imaging.

For phase-based conductivity imaging, a TSE scan has been acquired. The in-plane resolution was set to $1.1 \times 1.1\,\mathrm{mm}^2$ with 10 mm slice thickness. Other sequence parameters were $TE = 11.32\,\mathrm{ms}$, $TR = 600\,\mathrm{ms}$, TSE factor 6, leading to an overall imaging time of $5:05\,\mathrm{min}$. Conductivity reconstruction was performed using 3 voxels on each side of the targeted

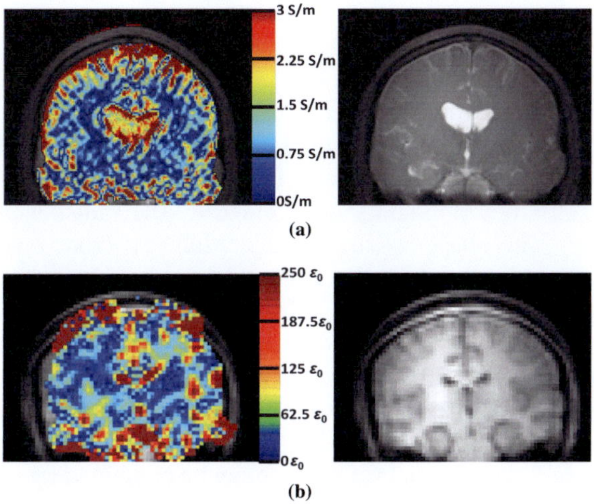

Figure 3.12: High field EPT conductivity and permittivity maps. (a) shows conductivity reconstruction with 1.1 mm in-plane resolution as well as the corresponding anatomy (acquired using a TSE sequence). The conductivity map was reconstructed based on phase only. (b) Permittivity map and corresponding anatomy (acquired using an FFE sequence).

voxel for in-plane derivatives and 1 voxel in through-plane direction. For this high spatial resolution, no B_p map could be obtained within a reasonable imaging time, and thus, only phase-based conductivity mapping was performed. Resulting conductivity images are shown in Fig. 3.12a. Clearly, a correlation between CSF in the anatomic image and high conductivity values is observed.

3.7 Discussion

In the presented work, a stable and reliable EPT reconstruction method is derived and applied to phantom and *in vivo* cases. For the first time, quantitative conductivity and permittivity values of human tissue are determined *in vivo* using standard MRI. The values obtained for GM, WM, and CSF agree well with corresponding literature values according to, e.g., [97, 107]. However, one has to keep in mind that literature values are frequently animal and / or *ex vivo* data, hampering a direct comparison with the human *in vivo* data presented in this study. The validity of the proposed method is confirmed in phantom experiments. Expected values could be matched within one standard deviation for conductivity and permittivity average values.

Phase-based conductivity imaging is introduced and successfully tested in simulation studies and in experiments on healthy volunteers. Conductivity contrast is predominantly determined by the transmit phase φ_p of the RF pulse. The phase-based approximation particularly works in the case of $\omega\varepsilon \ll \sigma$. Fortunately, this is the case for most human tissue types at 1.5 T, leading to good clinical applicability of phase-based conductivity imaging. Images reconstructed using the exact formula on the one hand, and the phase-based approximation on the other hand agree very well, also quantitatively. Errors introduced by skipping the lengthy B_p magnitude measurement were of the order of 10%. A bias towards higher values was found (see Tabs. 3.2, 3.3). This bias can be explained by the observed interplay of convex magnitude and concave phase distribution of the RF transmission field. The resulting anti-parallel gradients of B_p and φ_p yield a negative dot product in Eq. 3.17, and thus, a negative contribution of the neglected component in Eq. 3.19. With phase-based reconstruction, conductivity imaging was promoted to a stage, where clinical application seems feasible, since scan times could be drastically reduced. In this study, a 3D scan suitable for conductivity reconstruction with 1.6 mm in-plane and 2 mm through-plane resolution and 11 slices has been performed in 5 min. Thus, phase-based conductivity imaging seems to be mature for clinical evaluation.

In analogy, permittivity is predominantly determined by the magnitude B_p. Magnitude-based permittivity imaging particularly works for tissues with $\omega\varepsilon \gg \sigma$. Therefore, magnitude-based permittivity imaging is theoretically less suitable for imaging highly conductive body fluids like CSF or blood. This fact has been observed in simulation studies, where exact and approximate permittivity reconstructions diverged for CSF. For most other tissues, acceptable errors of the order of 20% were observed in simulation studies. In the low resolution *in vivo* study, hardly any voxel containing exclusively CSF could be found due to partial volume effects. Moreover, the applied calculus operations are defined on an ensemble of voxels containing also white matter adjacent to CSF. Therefore, errors introduced by magnitude-based reconstruction were found less severe than in the simulation study. Quantitative values from the exact reconstruction and magnitude-based approximation are similar and in good agreement with literature values. In general, magnitude-based reconstruction yields lower values than the exact approach (see Tabs. 3.2, 3.3), since the transition from Eq. 3.18 to Eq. 3.20 omits a component, which is quadratic, and thus, is always positive. SNR of reconstructed permittivity is significantly lower than SNR of reconstructed conductivity due to the following two reasons. 1) The impact of the permittivity on the amplitude B_p is much lower than the impact of the conductivity on the phase φ_p. 2) The noise present in B_p maps is typically much higher than the noise present in the image phase. Thus, longer imaging times are required, leading in the presented study to a scan time of 13 min for a 3D data set covering the whole head with 3.1 mm² in-plane and 15 mm through-plane resolution. In future work, methods to improve SNR and/or spatial resolution of permittivity imaging shall be developed.

In principle, the EPT method can be used in combination with any B_p/φ_p mapping method. The accuracy of EPT depends on the accuracy of the mapping, i.e., the most accurate B_p/φ_p mapping method leads to the most accurate EPT results. The accuracy of two different B_p mapping methods has been investigated in Chapter 2. MTM has been proven to yield good SNR per unit of measurement time at little systematic errors. In the future, also Bloch Siegert B_p mapping (Section 1.13) should be investigated for EPT purposes.

To assess the feasibility of phase- and magnitude-based approximations at B_0 higher than 1.5 T as in this study, several additional effects have to be taken into account. On the one hand, according to the 4-Cole-Cole model [97], conductivities increase and permittivities decrease with increasing B_0. On the other hand, also ω increases, leading to $\sigma \approx \omega\varepsilon$ at $B_0 = 2.3/3.7/14$ T for WM/GM/CSF, respectively as shown in Fig. 3.13. This implies a better applicability of magnitude-based permittivity imaging at high field strength, but simultaneously, hampering phase-based conductivity imaging. At 3 T however, no significant degradation of phase-based conductivity imaging could be observed. Increasing ω also leads to higher B_p magnitude variations, when the RF wavelength becomes comparable to patient dimensions, i.e., the impact of ε on B_p increases. This would improve the sensitivity of EPT for permittivity measurements as well as the approximation underlying Eq. 3.20 for magnitude-based per-

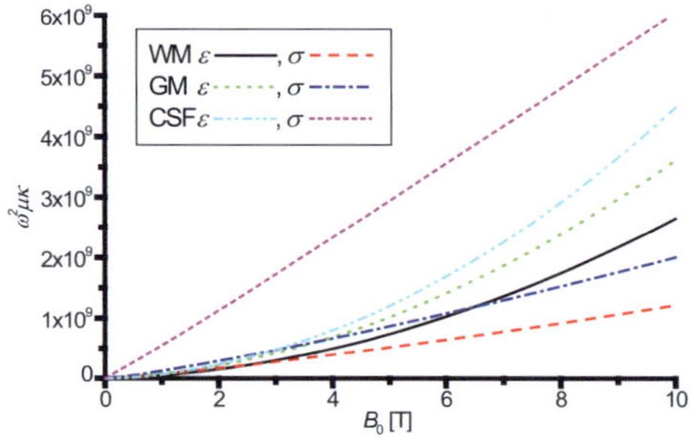

Figure 3.13: Dielectric properties of head tissue based on the 4-Cole-Cole formula as a function of main field strength. Conductivity values increase linearly with frequency, whereas permittivity increase is quadratic. This implies better imaging of dielectric properties at higher field strength, and being especially beneficial for permittivity imaging.

mittivity imaging. Thus, it is expected that the feasibility of full as well as magnitude-based permittivity imaging increases at higher field strength. This could be confirmed at least for $B_0 = 3\,\text{T}$, where permittivity maps with higher resolution at similar SNR could be obtained within comparable acquisition times.

In general, the dielectric properties κ are described by a tensor. Also in the human brain, preferred axon orientation in the white matter very likely leads to a directional dependence of conductivity and/or permittivity. However, anisotropy is frequency dependent and at least in muscle smaller than 1% at MR Larmor frequency of 64 MHz [97]. At 64 MHz, cell membranes are "invisible" to electromagnetic fields and intracellular cytoplasm determines the resistivity of tissue [101] leading to directionally independent dielectric properties. Thus, the present study does not account for anisotropy, but assumes isotropic dielectric properties.

3.8 Conclusion

EPT provides a suitable framework to investigate the electric properties of human tissue *in vivo* with a standard MR system. Using the methods presented in this study, particularly conductivity imaging of the brain seems to become clinically feasible. On the other hand, extending EPT to further anatomies remains a promising task.

3.9 Acknowledgments

The authors of [108] would like to thank Julia Bohnert and Sebastian Seitz for substantial help on visible human FDTD simulations as well as Hanno Homann and Christian Findeklee for fruitful discussion.

CHAPTER FOUR

PATIENT-SPECIFIC LOCAL SAR DETERMINATION: *In Vivo* MEASUREMENTS AND FDTD VALIDATION

Dans la vie, rien n´est à craindre, tout est à comprendre.

— MARIE SKŁODOWSKA-CURIE (1867 – 1934)

Abstract — Heating of tissue during MR measurements is a major problem at high-field MRI, and particularly, in the framework of parallel radiofrequency (RF) transmission. The heating is directly related to the RF energy absorbed during an MR examination, i.e., the specific absorption rate (SAR), requiring reliable SAR estimation methods. Currently used SAR estimation methods are based on models which are neither patient-specific nor take into account patient position and posture, which typically leads to unnecessary large safety margins. In this work, an approach is presented, which measures local SAR in a patient-specific manner. Using a specific formulation of Maxwell's equations, the local SAR can be estimated based on post-processing of the complex transmit sensitivity of the RF antenna involved. The approximations involved in the proposed method are investigated. The presented approach yields a sufficiently accurate and patient-specific local SAR measurement within a scan time of less than 5 min.

Based upon:

- Voigt T, Homann H, Doessel O, Katscher U. Patient-Individual Local SAR Determination: *In Vivo* Measurements and FDTD Validation. Under review in *Magnetic Resonance in Medicine*.
- Voigt T, Homann H, Katscher U, Doessel O. Patient-specific *in vivo* Local SAR Estimation and Validation. *In Proceedings of the 18th Annual Meeting of ISMRM*, Stockholm, Sweden 2010. p. 3876.

4.1 Introduction

During an MRI procedure, the patient absorbs a portion of the transmitted radiofrequency (RF) energy, which can result in tissue heating. The so-called specific absorption rate (SAR), in Watts per kilogram (W/kg), is the RF power absorbed per unit mass of tissue. Especially in high-field MRI, when the size of the body becomes comparable to the wave length of the applied RF field, standing wave patterns can lead to RF field inhomogeneities and may cause elevated local heating. Here, MRI using multiple transmit channels allows homogenizing transmitted RF fields, but also contains the enhanced risk of formation of so-called hot spots, i.e., local maxima of the SAR [84]. SAR, as one of the most important parameters related with thermal effects, acts as a guideline for MRI safety [13]. Legal regulations put upper limits to allowed local and global SAR applied during MR imaging [109]. To comply with these regulations, global and local SAR has to be estimated prior to scanning and the SAR relevant scan parameters adjusted to meet regulatory limits [110]. This has to be done before any MR examination, potentially limiting the parameter space available for the current MR sequence. Moreover, patients with metallic implants are usually excluded from MR scans due to possible RF heating of these devices, which could lead to tissue burning.

Typically, it is more likely to violate local SAR limits than global or partial body SAR limits [111, 112]. The determination of local SAR is therefore of central importance for high RF energy MR applications. Unfortunately, unlike whole body and partial body SAR, local SAR is not accessible via a straight-forward RF power-based estimation due to its dependence on the tissue's local dielectric properties and mass density as well as the local, complex three-dimensional (3D) electric field distribution. Thus, the problem of local SAR determination is often addressed in numerical field calculations using suitable coil and patient models. Particularly the finite-difference time-domain (FDTD) technique [113] is widely used to compute fields and local SAR. For instance, FDTD has been used to determine local SAR in a birdcage coil loaded with the human head [114–117], for local SAR calculations employing surface coils [111] and whole body coils [118], and applied to studies on local SAR in fetuses [119, 120]. Based on FDTD local SAR calculations, the resulting tissue heating has been investigated using thermal models [121, 122]. Moreover, FDTD results were found to agree well with experimentally determined B_p field distributions [36,123]. Typically, FDTD simulations require several hours of computation time, not including patient and coil model generation. Therefore, patient / position / posture-specific simulations to determine local SAR are usually too computationally demanding for clinical routine. Instead, simulation results of a single standard model subject, for instance a male from the visible human project [124], are used together with suitable safety margins. The dielectric properties of the visible human at MR Larmor frequency are usually determined from published data [97].

Besides being practicable, SAR simulations of a standardized subject show several disadvantages. Particularly, standardized subject SAR simulations suffer from limited accuracy [87], i.e. the actual local SAR pattern varies significantly with patient size [118] and patient position [125]. Another source of error can be a non-specific assignment of dielectric properties to the patient model. It has been shown that the exact value of electric properties is important for local SAR determination [126]. Thus, an exact determination of local SAR has to be based on patient-specific data. Furthermore, idealizing the RF transmit chain and coil can also be a source of error for SAR simulations. In real settings, imperfections or malfunctions along the transmit chain could lead to completely different RF and SAR pattern. Remaining model errors enforce large safety margins, and therefore, might lead to wasted imaging capabilities with respect to speed and / or SNR. An exact patient-specific determination of local SAR might help to fully leverage hardware capabilities.

Recently, direct local SAR measurements have been addressed in some publications. The determination of local SAR based on the RF fields applied during MR examination has been suggested by [63] in the framework of electric properties tomography (EPT) and applied preliminary to *in vivo* experiments [96]. Conceptual studies of B_p based SAR determination have also been performed in the framework of simulated data [88].

Since the local SAR is closely related to heat induction, [127] attempted to estimate the local SAR via temperature mapping. [72] used the phase difference method [128] to determine the temperature increase and linked this increase to the local SAR assuming a linear relation. In contrast, RF based methods like EPT to determine the local SAR have the advantage that no significant heating has to be induced to obtain local SAR maps. On the other hand, so far no validation of the acquired maps and no in depth investigation of *in vivo* feasibility of RF based local SAR mapping has been performed.

In this study, local SAR is determined by post-processing a combination of receive and transmit RF phases as encountered in standard spin echo imaging. This work can be seen as a further development of the approach presented in the EPT framework. Furthermore, the acquired local SAR maps are FDTD validated using a realistic coil model as well as a realistic patient model based on individual data. The presented approach yields a patient-specific local SAR determination with sufficient accuracy and within short computation time.

4.2 Theory

The aim of this section is to link the local SAR to quantities that are accessible via MRI. The local SAR is defined as

$$\mathrm{SAR}(\mathbf{r}) = \frac{\sigma(\mathbf{r})}{2\rho(\mathbf{r})} |\mathbf{E}(\mathbf{r})|^2 \tag{4.1}$$

with σ being the electric conductivity, ρ the mass density and $\mathbf{E}$ the electric field vector. Throughout this chapter, a constant mass density of tissue is assumed, i.e. $\rho(\mathbf{r}) = \rho$. For an otherwise exact determination of local SAR, a spatial distribution of the electric field of the involved RF coil throughout the patient as well as the electric conductivity distribution throughout the patient are required. To determine these quantities, a circularly polarized basis of the spatial coordinates is most suitable. The new basis vectors $\mathbf{e}_p$, $\mathbf{e}_m$, $\mathbf{e}_z$ are related to Cartesian $\mathbf{e}_x$, $\mathbf{e}_y$, $\mathbf{e}_{z,cart}$ via $\mathbf{e}_p = \frac{1}{\sqrt{2}}(\mathbf{e}_x + i\mathbf{e}_y)$, $\mathbf{e}_m = \frac{1}{\sqrt{2}}(\mathbf{e}_x - i\mathbf{e}_y)$, and $\mathbf{e}_z = \mathbf{e}_{z,cart}$, and thus, are mutually orthogonal. In these coordinates, the magnetizing field vector reads $\mathbf{H} = (H_p, H_m, H_z)^T$. The circular polarized components are related to the Cartesian components via $H_p = \frac{1}{\sqrt{2}}(H_x + iH_y)$, $H_m = \frac{1}{\sqrt{2}}(H_x - iH_y)$, and $H_z = H_{z,cart}$. The positively rotating component H_p couples to the proton spins, and thus, is accessible via MR. The dielectric properties conductivity σ and permittivity ε can be linked to this field via

$$\frac{\oint_{\partial V} \nabla H_p(\mathbf{r})\,\mathrm{d}\mathbf{a}}{\omega^2 \mu \int_V H_p(\mathbf{r})\,\mathrm{d}V} = \kappa(\mathbf{r}), \tag{4.2}$$

where the dielectric properties are summarized to $\kappa(\mathbf{r}) = \varepsilon(\mathbf{r}) - i\frac{\sigma(\mathbf{r})}{\omega}$. The derivation of Eq. 4.2 is provided in Section 3.3. The positively rotating component is a complex quantity $H_p = \mu^{-1} B_p \exp(i\varphi_p)$, where B_p represents the magnitude of the positively rotating component of the transmit magnetic field and φ_p its phase.[1] For separate conductivity and permittivity reconstruction, the following expressions can be obtained

$$\sigma(\mathbf{r}) = (\omega\mu V)^{-1} \left(\oint_{\partial V} \nabla\varphi_p(\mathbf{r})\, \mathrm{d}\mathbf{a} + 2\int_V \left[\nabla \ln\left(B_p(\mathbf{r})\right) \cdot \varphi_p(\mathbf{r})\right] \mathrm{d}V \right) \tag{4.3}$$

$$\varepsilon(\mathbf{r}) = \left(\omega^2 \mu \int_v B_p(\mathbf{r})\, \mathrm{d}V \right)^{-1} \left[\int_V \left(\nabla\varphi_p(\mathbf{r})\right)^2 B_p(\mathbf{r})\, \mathrm{d}V - \oint_{\partial V} \nabla B_p(\mathbf{r})\, \mathrm{d}\mathbf{a} \right]. \tag{4.4}$$

A detailed derivation of Eqs 4.3 and 4.4 is given in Section 3.3. The furthermore required electric fields $\mathbf{E}$ are not directly accessible in standard MR systems. However, the electric fields can be expressed as function of the magnetizing field vector $\mathbf{H}$ and dielectric properties κ via Ampere's law

$$\mathbf{E}(\mathbf{r}) = \frac{\nabla \times \mathbf{H}(\mathbf{r})}{i\omega\kappa(\mathbf{r})} \tag{4.5}$$

where time harmonic fields $\mathbf{E}, \mathbf{H} \propto \exp(i\omega t)$ have been assumed. However, to calculate $\mathbf{E}$ via Eq. 4.5, not only H_p is needed, but also H_m and H_z, which are not accessible experimentally. Thus, the aim of the following derivations is to investigate the impact of neglecting these unknown magnetizing field components.

First, Amperes law is rewritten using the above introduced basis $\mathbf{e}_p$, $\mathbf{e}_m$,$\mathbf{e}_z$. The change of basis $\Phi : \mathbb{C}^3 \to \mathbb{C}^3$ from Cartesian $\mathbf{r} = (x,y,z)^T$ to the circular coordinate system $\tilde{\mathbf{r}} = (p,m,z)^T$ can be described by the matrix $\mathbf{J}$ given in Eq. 3.5. Using this matrix, the coordinate transformation of position vector $\mathbf{r}$ is stated in Eq. 3.6. Further properties of this transformation are described in Section 3.3 of this work.

[1] $\mathbf{H}$ will be referred to as magnetizing field and $\mathbf{B}$ as magnetic field according to the notation of [102]. Since constant μ is assumed, $\mathbf{H}$ and $\mathbf{B}$ differ only by a constant scaling factor.

Ampere's law relates the electric field vector to the magnetizing field vector. Using $\nabla = \mathbf{J}^{\dagger}\tilde{\nabla}$, $\mathbf{H}(\mathbf{r}) = \mathbf{J}^{\dagger}\tilde{\mathbf{H}}(\mathbf{r})$, and $\mathbf{E}(\mathbf{r}) = \mathbf{J}^{\dagger}\tilde{\mathbf{E}}(\mathbf{r})$ yields

$$i\omega\kappa(\mathbf{r})\mathbf{J}^{\dagger}\tilde{\mathbf{E}}(\mathbf{r}) = \mathbf{J}^{\dagger}\tilde{\nabla}\times\mathbf{J}^{\dagger}\tilde{\mathbf{H}}(\mathbf{r}) \tag{4.6}$$

where the tilde denotes circular representation. The transformation matrix can be removed from the cross product according to

$$i\omega\kappa(\mathbf{r})\mathbf{J}^{\dagger}\tilde{\mathbf{E}}(\mathbf{r}) = \det\left(\mathbf{J}^{\dagger}\right)\left(\mathbf{J}^{\dagger}\right)^{-T}\tilde{\nabla}\times\tilde{\mathbf{H}}(\mathbf{r}). \tag{4.7}$$

Multiplication by $\mathbf{J}$ from the left hand side yields

$$i\omega\kappa(\mathbf{r})\mathbf{J}\mathbf{J}^{\dagger}\tilde{\mathbf{E}}(\mathbf{r}) = i\mathbf{J}\mathbf{J}^{T}\tilde{\nabla}\times\tilde{\mathbf{H}}(\mathbf{r}). \tag{4.8}$$

The matrix product is evaluated according to

$$i\omega\kappa(\mathbf{r})\tilde{\mathbf{E}}(\mathbf{r}) = i\begin{pmatrix} 0 & 1 & 0 \\ 1 & 0 & 0 \\ 0 & 0 & 1 \end{pmatrix}\tilde{\nabla}\times\tilde{\mathbf{H}}(\mathbf{r}). \tag{4.9}$$

Thus, the electric field vector in circular coordinates can be written as

$$\begin{aligned} i\omega\kappa(\mathbf{r})E_p(\mathbf{r}) &= i\left(\partial_z H_p(\mathbf{r}) - \partial_p H_z(\mathbf{r})\right) \\ i\omega\kappa(\mathbf{r})E_m(\mathbf{r})(\mathbf{r}) &= i\left(\partial_m H_z(\mathbf{r}) - \partial_z H_m(\mathbf{r})\right) \\ i\omega\kappa(\mathbf{r})E_z(\mathbf{r})(\mathbf{r}) &= i\left(\partial_p H_m(\mathbf{r}) - \partial_m H_p(\mathbf{r})\right). \end{aligned} \tag{4.10}$$

Using the approximations $H_z = 0$ and $H_m = 0$ yields

$$\begin{aligned} i\omega\kappa(\mathbf{r})E_p(\mathbf{r}) &= i\partial_z H_p(\mathbf{r}) \\ i\omega\kappa(\mathbf{r})E_m(\mathbf{r}) &= 0 \\ i\omega\kappa(\mathbf{r})E_z(\mathbf{r}) &= -i\partial_m H_p(\mathbf{r}). \end{aligned} \tag{4.11}$$

In contrast to intuition, neglecting H_m and H_z is equivalent to assuming zero negatively rotating electric field and approximating the other two components. In this approximation, the magnitude of the electric field can be written as a function of the positively rotating component of the magnetizing field

$$\begin{aligned}|\mathbf{E}(\mathbf{r})|^2 &= \left|\tilde{\mathbf{E}}(\mathbf{r})\right|^2 = |\omega\kappa(\mathbf{r})|^{-2}\left[\left|\partial_z H_p(\mathbf{r})\right|^2 + \left|\partial_m H_p(\mathbf{r})\right|^2\right] \\ &= |\omega\kappa(\mathbf{r})|^{-2}\left[\left|\partial_z H_p(\mathbf{r})\right|^2 + \frac{1}{2}\left|\partial_x H_p(\mathbf{r})\right|^2\right. \\ &\quad \left. + \frac{1}{2}\left|\partial_y H_p(\mathbf{r})\right|^2 + \Re\left(i\partial_x H_p(\mathbf{r})\,\partial_y H_p(\mathbf{r})^*\right)\right]. \qquad (4.12)\end{aligned}$$

The local SAR can now be determined based on H_p according to Eq. 4.1 using Eq. 4.2 for the dielectric properties and Eq. 4.12 for the electric fields. Due to its approximate nature, the validity of the latter expression Eq. 4.12 has to be proven. This is done in the next Section 4.3.

In Chapter 3, one of the main findings was that eddy current effects are predominantly expressed in MR image phase. Hence, a conductivity estimation based on image phase was introduced and successfully tested. This motivates to estimate also the local SAR via phase information only. Assuming a constant amplitude B_p in Eq. 4.3, conductivity can be expressed as a function of transmit field phase [108]

$$\sigma(\mathbf{r}) = (\omega\mu V)^{-1}\left(\oint_{\partial V} \nabla\varphi_p(\mathbf{r})\,\mathrm{d}\mathbf{a}\right). \qquad (4.13)$$

Again, a thorough motivation and testing of this approximation is given in Chapter 3. Setting $B_p = const$ and using $H_p = \mu^{-1}B_p\exp(i\varphi_p)$ in Eq. 4.12 yields an expression for the electric fields in Eq. 4.1, based on transmit phase variations

$$|\mathbf{E}(\mathbf{r})|^2 = |\omega\kappa(\mathbf{r})|^{-2}B_p^2\left[\frac{1}{2}(\partial_x\varphi_p)^2 + \frac{1}{2}(\partial_y\varphi_p)^2 + (\partial_z\varphi_p)^2\right]. \qquad (4.14)$$

In the formalism described above, not only the conductivity, but also the permittivity enters the SAR determination via Eq. 4.5. However, using phase data only, the permittivity cannot be approximated with EPT and needs to be determined in a different manner. In this work, alternative ways to determine the local SAR will be investigated:

1. Determine the local SAR using the measured complex H_p field according to Eqs. 4.12 and 4.1, and the dielectric properties according to Eq. 4.2.

2. Determine the local SAR using the measured complex H_p field according to Eqs. 4.12 and 4.1, and the dielectric properties are obtained from a look-up table [97] and assigned to a segmented patient geometry. This approach is called the hybrid approach.

3. Determine the local SAR and conductivity based only on phase imaging only according to Eq. 4.13 and 4.14. Permittivity is assumed to be constant, $\varepsilon = 80$.

The mass density is assumed to be constant ($\rho = 1000\,\mathrm{kg/m^3}$) in all cases. According to [118], densities for brain tissue are $\rho_{\mathrm{GM}} = 1038\,\mathrm{kg/m^3}$, $\rho_{\mathrm{WM}} = 1038\,\mathrm{kg/m^3}$, and $\rho_{\mathrm{CSF}} = 1007\,\mathrm{kg/m^3}$.

4.2.1 Magnetic field magnitude and phase imaging

The presented approach for SAR measurements applies the equations discussed above to the RF field generated by a transmit (Tx) / receive (Rx) RF coil of a standard MR system. The magnitude of the spin tipping component H_p can be measured via corresponding mapping techniques (see, e.g., Chapter 2 of this work). Usually, the B_p map is acquired as the percentage of the nominal flip angle α_{nom} preset at the scanner console. To estimate absolute values of the local SAR, a magnetic field value in units of Tesla is necessary. Assuming a hard pulse it is

$$B_p(\mathbf{r}) = B_{p,rel}(\mathbf{r})\,\frac{\alpha_{nom}}{\tau\gamma} \tag{4.15}$$

where γ is the gyromagnetic constant and τ is the duration of the pulse. Equation 4.15 allows using low SAR (i.e., low flip angle) B_p maps to estimate the absolute scaling of B_p of any arbitrary pulse with α_{nom} and τ. From known local SAR distributions, different pulse shapes, pulse durations, and repetition times can be taken into account according to [114, 129].

As discussed in the last paragraph, the knowledge of the Tx phase φ_p is required for local SAR determination. No method has been found in the literature to exactly determine φ_p. The signal phase of an MR image is given by $\varphi = \varphi_p + \varphi_m'$ and contains both, the Tx phase φ_p and the Rx phase φ_m' (see also Section 3.3.4). The prime indicates that the coil used for reception is not necessarily identical (or identically driven) with the coil used for transmission. In this study, these two phases are separated by the approximation $\varphi_p \approx \varphi/2$. This relation is exactly valid in an empty quadrature volume coil with a switch of polarization between Tx and Rx, and turned out to be a good approximation if such a coil is loaded at 64 MHz [63, 95] as well as higher field strengths [104].

Using different coils for transmission and reception will lead to contributions from the sensitivities of both coils to φ. However, as explained in Section 3.3.4, this feature does not impact the reconstructed conductivity as long as the phase-based assumption, and thus, the linearity of the reconstruction is valid. Local SAR imaging as presented in this chapter requires exact knowledge of φ_p. The Rx phase contribution of a deticated Rx coil might be removed with the help of a concept called CLEAR.[2] For CLEAR, the ratio between surface coil and body coil images is measured in an additional preparation scan. The corresponding CLEAR reconstruction replaces the sensitivity of the applied Rx coil by the Rx sensitivity of the (switched polarization) body coil, hence allowing the above mentioned approximation, i.e. dividing the image phase by two. The impact of a CLEAR reconstruction on phase imaging is investigated experimentally in the next section using phantoms.

The presented SAR imaging method is based on imaging the effects of eddy currents arising from the transmitted RF pulse. Any changing magnetic flux induces eddy currents which influence B_p magnitude and φ_p phase images. Thus, another influence on the image is expected from eddy currents arising from gradient switching. It has to be made sure that those currents arising from other sources than the RF pulse are removed from the images to ensure that H_p is measured correctly. In this work, eddy currents induced by the read gradient are removed by averaging two subsequent SE measurements with inverted polarity of the read gradient [130].

B_0 inhomogeneities are not expected to play a major role in phase imaging using SE sequences, since the phase caused by off resonance effects is assumed to be refocused by the 180° pulse.

[2]On the system used in this study, a CLEAR reconstruction is defined as a SENSE reconstruction [57] with reduction factor one.

4.3 Methods

4.3.1 FDTD simulations

FDTD simulations have been conducted to test the impact of the approximations made in the theory section on the local SAR (Semcad, Schmid & Partner Engineering AG, Zürich, Switzerland). The head and the shoulders of the visible human were placed inside a quadrature body coil (Fig. 4.1a). The coil was driven in resonant mode and tuned to 64 MHz. The electric properties of the head were set according to [97]. A resolution of $1 \times 1\,\text{mm}^2$ in-plane was used to simulate all magnetic field components. The local SAR was reconstructed using:

1. The exact formula for local SAR taking into account all magnetic field components (Eq. 4.1).
2. Using the exact formula but neglecting the H_z component.
3. Reconstruct local SAR based on the complex H_p field only (Eq. 4.12), i.e., neglecting H_z and H_m.
4. Reconstruct local SAR using only the transmit phase and assuming constant B_p (Eqs. 4.13 and 4.14).

The resulting SAR images were compared using a correlation value inside a region of interest (ROI) comprising 21000 pixels. Additionally, the 10 g average local SAR and the global SAR were calculated over the whole field of view according to [131] and compared for all cases investigated. The applied (simulated) RF pulse had an average B_p amplitude of $3.1\,\mu\text{T}$ over the whole volume of interest, which corresponds to a 90° flip angle at RF duration of approximately 2 ms (cf. Eq. 1.8). For computation of the derivatives in the mathematical expressions, Savitzky-Golay filtering [106] was used, i.e., a set of convolution coefficients, obtained by a local polynomial regression, was applied to the simulated magnitude and/or phase maps. In this study, the convolution coefficients were based on a one dimensional parabola fitted to $I = 2$ voxels on each side of the targeted voxel for both in-plane directions and in through-plane direction. For computation of the integrals, $J = 1$ adjacent voxel was used in each direction.

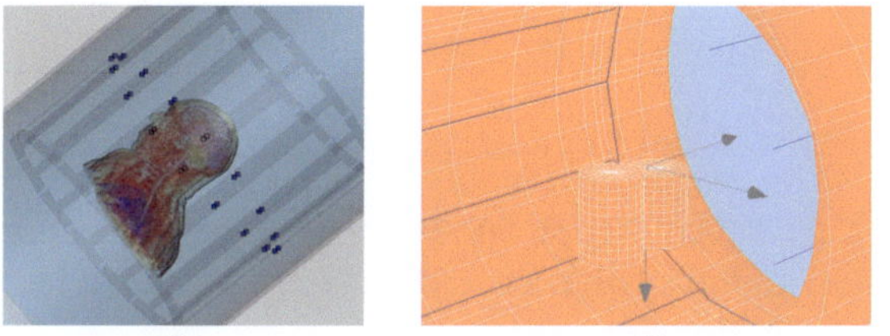

(a) (b)

Figure 4.1: Simulation and experimental setup used in this study. (a) Simulation setup used to test the applicability of the proposed method for local SAR imaging in the head. (b) Simulation of the imaging experiment using two cylinders with known conductivity and permittivity.

4.3.2 Phantom experiments

Phantom experiments are conducted to demonstrate the feasibility and accuracy of the proposed method. Two cylinders named A, B (both with diameter 7.5 cm and height 9 cm) filled with saline solution with known conductivity and permittivity were imaged using a standard 1.5 T scanner (Philips Healthcare, Best, The Netherlands). Electric conductivity values were determined with a 4-ring potentiometric probe (HI8733, Hanna Instruments, USA) and permittivity was determined using different mixing ratios of water and 2-propanol, yielding ($\sigma_1 = 0.42\,\mathrm{S/m}$, $\varepsilon_1 = 50$) for cylinder A and ($\sigma_2 = 1.35\,\mathrm{S/m}$, $\varepsilon_2 = 75$) for cylinder B. T_1 of the phantoms was reduced to approximately 300 ms using 2.5 ml Magnevist (Bayer Schering Pharma AG, Berlin, Germany) per liter of phantom fluid. The experimental setup is shown in Fig. 4.1b. Both cylinders were placed in a 8 element Rx head coil with a feet-head distance of 0.5 cm. The isocenter was 0.5 cm above the cylinders. The described setup was duplicated in a simulation environment. The simulation tool CONCEPT II (Technical University Hamburg–Harburg, Department of Theoretical Electrical Engineering, Germany) was used to compute magnetic field components [105]. The measured dielectric properties of the cylinders were used as input values for the simulation.

In a first experiment assessing the principle ability to measure transmit phase φ_p, an SE sequence was used with TR = 1500 ms, $\alpha = 90°$, 1.5×1.5 mm^2 in-plane resolution and 5 mm slice thickness. In all cases tested, the quadrature body coil (QBC) was used for RF transmission. In subsequent measurements, the following five different phase imaging / reconstruction techniques have been tested.

i. Using the head coil for reception in quadrature mode.

ii. Using the head coil for reception and perform CLEAR image reconstruction. To this goal, a standard low resolution reference scan was acquired additionally.

iii. Using the body coil for reception. Two averages have been acquired in this case to increase SNR.

iv. Using the head coil for reception and perform CLEAR image reconstruction. This time, a customized reference scan with the same (high) resolution as the SE scan ($1.5 \times 1.5 \times 5\,\mathrm{mm}^3$) has been acquired.

v. Using the head coil for reception and CLEAR reconstruction based on a high-resolution reference scan as in iv. Additionally, eddy current correction was performed. To perform the eddy current correction, the SENSE reference scan and the SE scan were repeated with inverted read gradient. The resulting images were averaged.

The resulting phases from cases i to v were multiplied with 0.5 to separate Tx from Rx phase and were compared to the simulated phase via calculating the corresponding correlation inside a circular ROI for each slice.

Using the same phantom and the same setup (Fig. 4.1), a second study was performed to demonstrate the feasibility and accuracy of the proposed local SAR imaging. For B_p mapping, the multiple TR B_p/T_1 mapping (MTM) sequence (Chapter 3) was applied using $\mathrm{TR}_{11,12} = 30/80\,\mathrm{ms}$, $\mathrm{TR}_{21,22} = 30/130\,\mathrm{ms}$, $\mathrm{TR}_{31,32} = 30/185\,\mathrm{ms}$, and $\mathrm{TR}_{41,42} = 30/200\,\mathrm{ms}$ with nominal flip angle 70° and total acquisition time of 23 min. For φ_p mapping, the phase of a 90° SE image was used as discussed in point v above. Over the entire FOV, an average B_p of $9.4\,\mu\mathrm{T}$ in the upper and $9.1\,\mu\mathrm{T}$ in the lower cylinder was measured according to Eq. 4.15.

The voxel size for all scans was set to $1.5 \times 1.5 \times 5\,\mathrm{mm}^3$. Conductivity reconstruction was performed according to Eq. 4.3 using a kernel of $I = 5$ pixels for in-plane derivatives and $I = 4$ pixels for through-plane derivatives. Permittivity reconstruction was performed according to Eq.

4.4 using $I = 8$ pixels in-plane and $I = 6$ pixels through-plane. For the integral, $J = 1$ pixel was used in both cases. Using these dielectric properties maps, the local SAR was calculated according to the H_p based Eq. 4.12 and using the phase-based approximation Eqs. 4.13 and 4.14.

The same setup was duplicated in a simulation environment. Local SAR was calculated using simulated electric fields (acc. to Eq. 4.1) and neglecting H_z and based on complex H_p only (acc. to Eq. 4.12). A correlation to the exact SAR image and the maximum 10 g average local SAR were computed for every case. The RF power applied in the simulations was adjusted to yield equality in simulated and experimental average B_p amplitude of $9.4\,\mu$T in the upper and $9.1\,\mu$T in the lower cylinder. The simulation results of the exact reconstruction and the reconstruction neglecting H_z can now be used to assess the error made in local SAR measurement using the proposed method.

4.3.3 Volunteer measurements

Coronal images of a healthy volunteer have been obtained using a SENSE Rx head coil with 8 channels in a 1.5 T MR scanner (Philips Healthcare, Best, The Netherlands). The head was covered with 50 slices and a spatial resolution of $5 \times 5 \times 5\,\text{mm}^3$. A B_p map was obtained in 3D acquisition mode using the MTM sequence (Chapter 2, [60]) with repetition times $\text{TR}_{11,12} = 30/320\,\text{ms}$, $\text{TR}_{21,22} = 30/200\,\text{ms}$, and $\text{TR}_{31,32} = 30/230\,\text{ms}$, leading to a total imaging time of $4:23+2:53+2:33\,\text{min} = 9:49\,\text{min}$. The option of overcontiguous slice acquisition has been enabled, reducing the actual through-plane resolution to 10 mm. A nominal flip angle of $\alpha = 65°$ and echo time TE = 1.43 ms was chosen. According to Eq. 4.15, the average B_p over the entire FOV was $9.4\,\mu$T.

For φ_p phase mapping, an SE sequence with the same geometry and TE = 6.5 ms / TR = 1500 ms / $\alpha = 90°$ was used. The multi-slice technique was used, i.e. the acquisition of k-space lines was performed in an interleaved fashion. The scan was repeated twice with inverted polarity of the gradients as discussed in the previous section, point v. A CLEAR reconstruction was performed on both scans and the results averaged and divided by two to yield the transmit phase estimate.

An additional scan of the same volunteer was performed to obtain a basis for tissue segmentation and model generation for simulation studies.

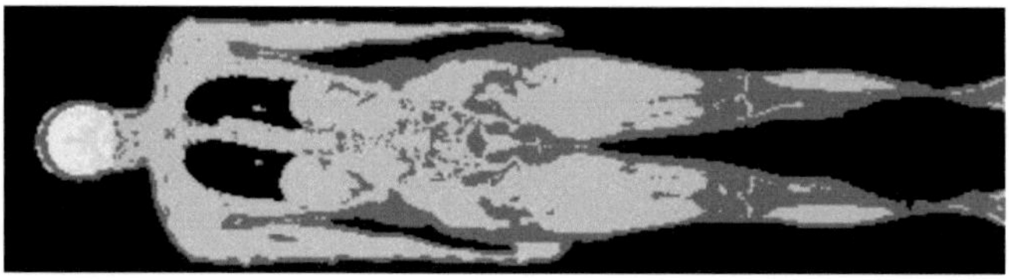

Figure 4.2: Dielectric model of the healthy volunteer created from Dixon water-fat scan. Muscle and fat have been segmented. In the head, also WM, GM, and CSF are separated. The resolution is $5 \times 5 \times 5\,\text{mm}^3$.

To this goal, a Dixon water-fat scan [132] was acquired using the parameters TR $= 7.4\,\text{ms}$, TE $= 2.6\,\text{ms}$, ΔTE $= 1\,\text{ms}$, $\alpha = 35°$, and a spatial resolution of $5 \times 5 \times 5\,\text{mm}^3$. The scan was applied several times using a stepwise moving table acquisition covering the whole body. A model of the individual volunteer was created by separating water and fat and assigning dielectric properties $\sigma_{fat} = 0.04\,\text{S/m}$, $\varepsilon_{fat} = 6.5$ and $\varepsilon_{muscle} = 72.3$, $\sigma_{muscle} = 0.69\,\text{S/m}$. In the head, gray matter, white matter, and CSF were segmented additionally. The resulting model is shown in Fig. 4.2. In separate studies, two tissue types and 5 mm resolution were found to sufficiently reproduce 10 g averaged local SAR values [133]. FDTD simulations using the described patient model together with a suitable QBC model have been used to determine magnetic fields and local SAR. The RF power applied in the simulations was adjusted to yield the same average B_p as found in the measurement.

The measured fields were used for conductivity reconstruction according to Eq. 4.3 using a kernel of $I = 3$ voxels for in-plane and for through-plane derivatives. Permittivity reconstruction was performed according to Eq. 4.4 using $I = 5$ voxels in all directions. For the integral, $J = 1$ voxel was used in both cases. Using these dielectric properties maps, the local SAR was calculated according to the H_p based Eq. 4.12.

To increase image SNR of reconstructed SAR, a hybrid approach was applied combining literature values according to [97] and Eq. 4.12 to compute electric fields. These fields are then used in Eq. 4.1 to compute local SAR.

Additionally, the local SAR has been reconstructed assuming constant B_p field using the phase-based approximation Eq. 4.13 for conductivity reconstruction and Eq. 4.14 for reconstruction of the electric fields. In this case, the permittivity was assumed to be constant $\varepsilon = 80$.

	max 10 g average local SAR	global SAR	image correlation
exact reconstruction	0.62 W/kg	0.188 W/kg	
neglecting H_z	0.60 W/kg	0.177 W/kg	98 %
based on complex H_p	0.46 W/kg	0.128 W/kg	91 %
phase-based	0.36 W/kg	0.098 W/kg	80 %

Table 4.1: Local SAR values relevant for regulatory limits and image correlation. Neglecting non-measurable magnetizing field components reduces the maximum local SAR and the global SAR values. On the other hand, the image correlation remains high, indicating that local SAR patterns are not distorted but only scaled globally.

4.4 Results

4.4.1 FDTD simulations

Results from FDTD simulations are shown in Fig 4.3. Fig.4.3 shows the local SAR calculation using all field components. The maximum 10 g local SAR average value using the exact reconstruction was 0.62 W/kg (see Tab. 4.1 for SAR values relevant for regulatory limits). Neglecting the H_z component yielded a comparable hot spot strength. Further neglecting the negatively polarized field component H_m in the SAR reconstruction, however, leads to a decrease of hot spot intensity by 25 %. In the case of phase-based SAR reconstruction, the decrease was even higher (40 %). A similar decrease was observed for the global SAR, were values decreased with neglecting field components in the reconstruction.

On the other hand, hot spots in all cases were not shifted more than 5 mm. Hence, the spatial correlations between approximated and exact SAR were very high for all cases investigated. Therefore, a head local SAR reconstruction at 1.5 T based on accurately measured complex H_p field should yield the correct hot spot location, however underestimating the intensity by 25 %.

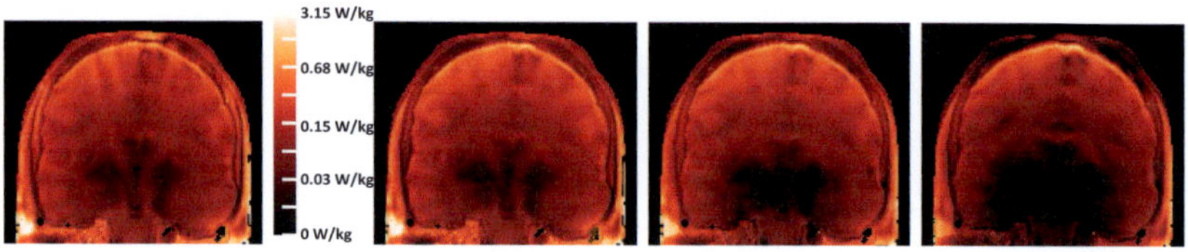

(a) (b) (c) (d)

Figure 4.3: Local SAR images from FDTD simulation. The impact of neglecting non-measurable magnetizing field components is shown. (a) is the exact local SAR reconstruction according to Eq. 4.1. (b) shows the local SAR reconstruction neglecting the z-component of the field. In (c), only the complex H_p field has been used for SAR reconstruction (Eq. 4.12). Reconstruction (d) is based on the transmit phase only, assuming constant B_p amplitude (Eq. 4.14).

4.4.2 Phantom experiments

To assess the ability of imaging the transmit phase φ_p, measured and simulated phase images are compared in Fig. 4.4. In all cases, the QBC has been used for transmission. The lowest correlation between measured and simulated transmit phase is obtained, if the quadrature head coil is used for signal reception in the experiment. The receive sensitivity of this coil can be removed using a CLEAR image reconstruction, and consequently, the correlation increases to a level of 90 % for all slices. As discussed in Section 4.2.1, the CLEAR reconstruction involves a usually low resolution reference scan. Using a high-resolution reference scan instead yields a slight further increase of the correlation. Using the quadrature body coil for reception yields less SNR as expected, however comparable accuracy even without CLEAR. Finally, a correlation of more than 95 % has been achieved by additionally using the described eddy current correction.

Electric properties reconstructed using B_p and φ_p are shown in Fig. 4.5. Apart from a boundary artifacts, the reconstructed permittivity and conductivity values are in agreement with the *a priori* determined values.

The results of the phantom local SAR imaging experiments are shown in Fig. 4.6. Key quantitative SAR parameters are shown in Tab. 4.2. The maximum 10 g average local SAR obtained from simulation was 1.44 W/kg in the exact reconstruction case. Neglecting H_z in the simulation led to a decrease of 26 %. In this phantom experiment, the influence of neglecting H_z is more significant than neglecting H_m in contrast to the FDTD study of the visible human, where the H_m component was found to be more relevant

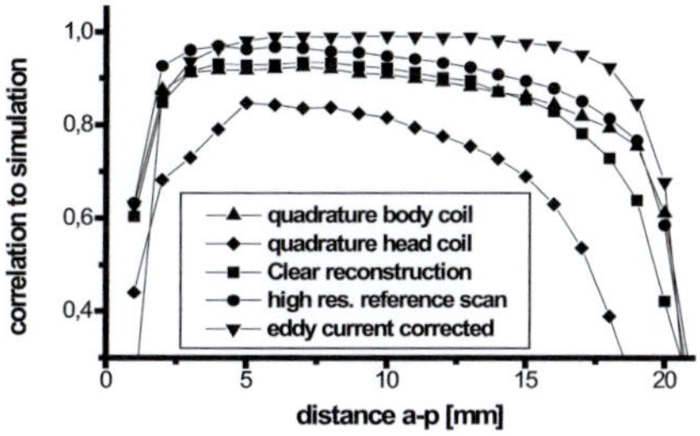

Figure 4.4: Comparison of measured and simulated transmit phase inside cylinder B. A correlation of 100 % indicates equality of the two phases. The best correlation has been obtained using CLEAR reconstruction and eddy current correction.

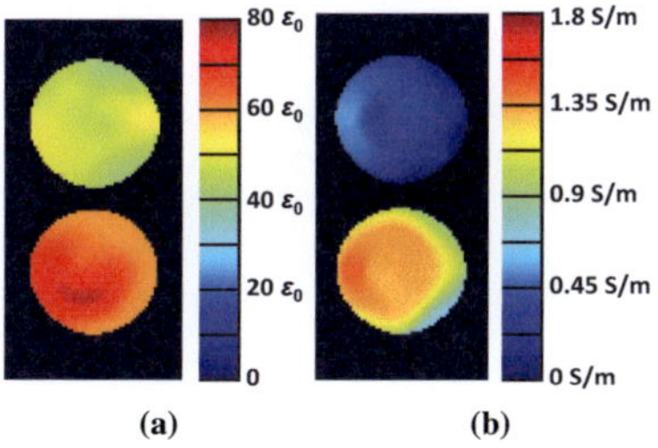

Figure 4.5: Electric properties determined using Eq. 4.2. (a) shows permittivity images of cylinder A (upper) and B (lower), conductivity is displayed in (b).

	max 10 g average local SAR	global SAR	image correlation
exact reconstruction	1.44 W/kg	0.48 W/kg	
neglecting H_z	1.06 W/kg	0.40 W/kg	98 %
based on complex H_p	1.04 W/kg	0.31 W/kg	97 %
phase-based	1.09 W/kg	0.29 W/kg	97 %

Table 4.2: Local SAR values relevant for regulatory limits and image correlation. Similar to the FDTD study on the visible human, the phantom simulation predict decreasing maximum 10 g average and global SAR values when neglecting field components in the SAR calculation.

(Chapter 4.4.1). Additionally neglecting H_m, i.e., using only the complex H_p for local SAR reconstruction had little influence on hot spot strength. This is also valid for measured local SAR, since the simulated fields have been scaled accordingly. The image correlation to the correct local SAR image was above 95 % for all cases investigated (see also Fig. 4.6). Hence, local SAR measurement using the presented approach yields accurate hot spot location and SAR pattern, however, lower overall SAR intensity. The findings from FDTD simulations on humans also hold for the cylindrical phantoms investigated.

4.4.3 Volunteer experiments

The results of the volunteer experiments are shown in Figs. 4.9 to 4.10. All steps from measurement to reconstruction have been validated with FDTD simulations using a dielectric model of the same individual. The results of H_p amplitude and phase mapping are shown in Figs. 4.8 and 4.7. Excellent agreement between measured and simulated amplitude and phase is observed along the profile in right-left direction; in the feet-head direction, minor deviations are visible in the lower part of the FOV (Figs. 4.8b and 4.7b). This is also visible in the corresponding images (Figs. 4.8a and 4.7a). Deviations could be due to measurement errors in B_p and / or φ_p mapping or due to the fact that the simple two-compartment (i.e., water/fat) segmentation, as applied in this area, might not be sufficient

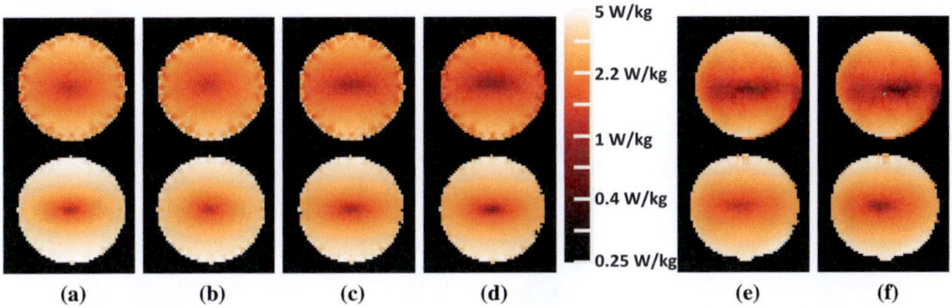

Figure 4.6: Phantom experiments. Simulation results are shown in (a) to (d), measured SAR distributions in (e) and (f). (a) Exact local SAR distribution. (b) Local SAR neglecting H_z. (c) Local SAR based on complex H_p field. (d) Local SAR estimation using the transmit phase φ_p only. (e) Measured local SAR distribution reconstructed according to Eq. 4.12. (f) Measured local SAR reconstructed from transmit phase only (Eq. 4.14).

to exactly reproduce the real fields. Please be reminded that only for the brain, additional compartments were taken into account (WM, GM, and CSF). MTM B_p mapping capability is additionally validated in Section 2.3.3.1.

Dielectric properties reconstructed according to Eq. 4.2 are shown in Fig. 4.9. Measured values for conductivity and permittivity are in the physiological range as expected from [134] and as found in the experiments presented in Chapter 3 of this thesis. Conductivity and permittivity spatially correlate with anatomy (Fig 4.9a).

Measured and simulated local SAR is shown in Fig. 4.10. The images reveal similar overall structure, a minimum in the central region as well as increasing local SAR towards the surface of the head and neck. Some differences between the local SAR images obtained from different reconstruction techniques are observed. Permittivity values reconstructed via EPT near the skull were higher than those stated in [97] (cf. Fig. 4.9c). Hence, using the hybrid approach with smaller $|\kappa|$ yields higher local SAR in that region (Fig. 4.10b). As expected from the simulation study, local SAR values using the phase-based reconstruction are lower than in all other cases (Fig. 4.10c). Comparing reconstructed local SAR with the simulation result, some voxels at the outer layer are lost due to the need of at least one (preferably more than one) voxel next to the reconstructed voxel. The maximum 10 g average SAR values are given in the caption

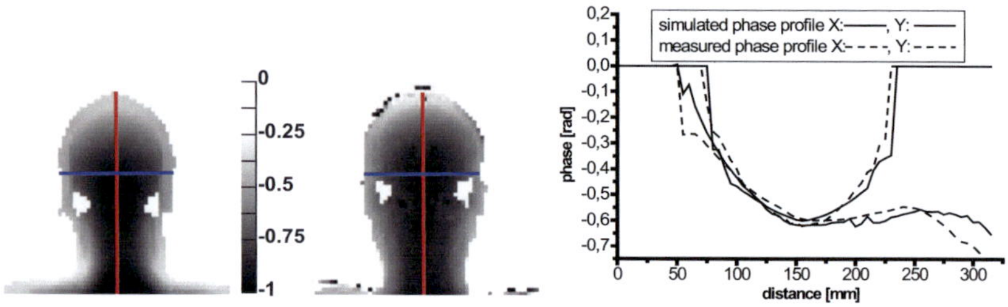

(a) (b)

Figure 4.7: Comparison of measured and simulated transmit phase. (a) shows a simulated phase image based on FDTD simulation of a dielectric model of the same individual on the left and the corresponding measured phase via SE sequence on the right. (b) shows phase profiles along the red and blue lines as indicated in (a).

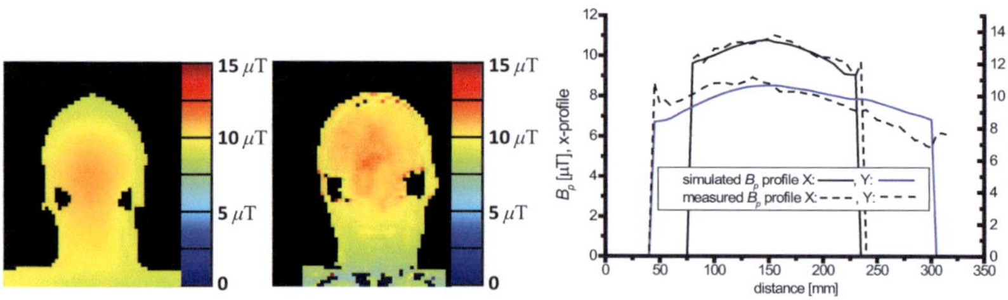

(a) (b)

Figure 4.8: Comparison of measured and simulated B_p maps. (a) shows simulated B_p on the left and measured B_p on the right. Besides slight deviations, the maps reveal similar overall structure. Particularly around the neck, simulated B_p is higher than measured in the experiment. The profiles shown in (b) are obtained at the same locations as in Fig. 4.7.

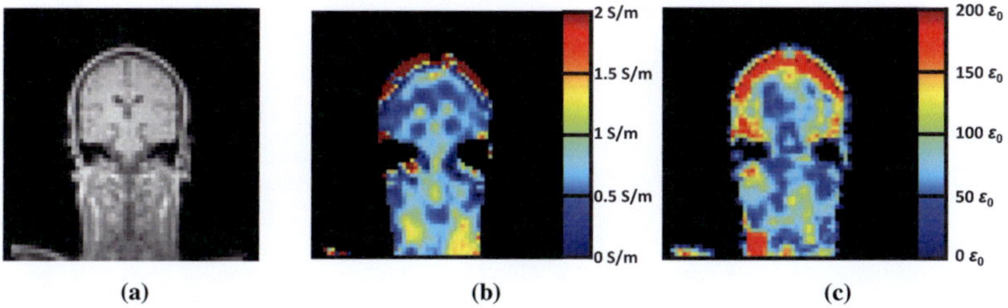

Figure 4.9: Intermediate images for SAR calculation. (a) Standard FFE image as by-product of MTM sequence. (b) Conductivity and (c) permittivity have been calculated using Eq. 4.3.

of Fig. 4.10 and 25 %-55 % lower than the value obtained from FDTD as expected from the visible human calculations. Please note that stated SAR values represent peak values on the basis of Eq. 4.15. Thus, the actual pulse shape did not enter the SAR estimation, nor did the repetition time (corresponding to a duty cycle of 100%). Taking these parameters into account, regulatory relevant maximum local SAR would be significantly lower.

4.5 Discussion

In this study, a new method to measure local SAR is presented and tested in simulations, phantom experiments, and volunteer measurements. EPT local SAR imaging estimates the local SAR from the active component H_p of the RF pulse. The link between the wanted quantity of interest, local SAR, and the measurable RF field is achieved via a suitable arrangement of Maxwell's equations of electromagnetism. In contrast to the gold standard of numerically simulating realistic, standardized patient models, this method yields a patient-specific local SAR estimate based on individual measurements. The approach assumes that the SAR contribution from the remaining two non-measurable magnetic field components is not significant. This assumption was found to be partially valid since it caused minor impact on the spatial structure of local SAR, particularly the hot spot positions. However, neglecting the non-measurable field components caused an overall local and global SAR reduction as expected from neglecting ad-

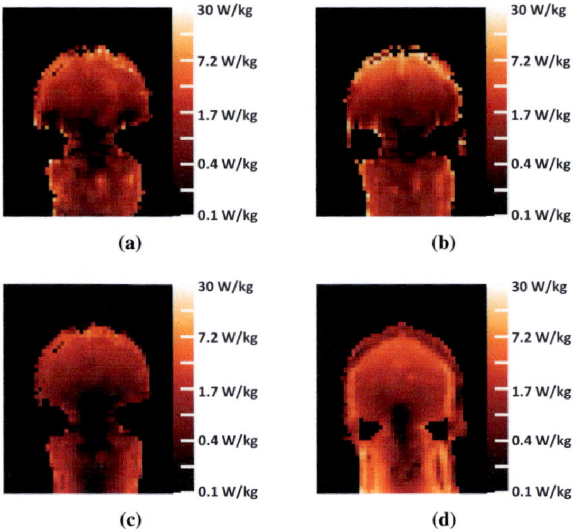

Figure 4.10: Local SAR images are shown in logarithmic scale. (a) shown local SAR reconstruction based on measured H_p, (b) is based on dielectric properties according to [97] (hybrid approach), (c) is the phase-based reconstruction, and the simulation result is shown in (d). The 10 g average maximum local SAR was 12.5 W/kg in (a), 15.6 W/kg in (b), 9.2 W/kg in (c), and 21.1 W/kg in (d). The simulation is based on a segmented model of the same individual.

ditive electric field contributions in Eq. 4.11. The reduction was found to be in the range of $30-40\,\%$ and affected the strength of hot spots and global SAR. Hence, the presented method is more suitable for hot spot detection and localization, which is of particular interest since local SAR is the limiting factor at high-field MRI. Typically, global SAR does not play such a critical role and is less relevant with regard to regulatory limits.

In brain studies, additional errors introduced by the applied assumption of a constant tissue density are only $3\,\%$ [118], reflecting the difference between GM/WM and CSF, and thus, negligible. When imaging other parts of the body, this error might increase, and a spatially varying tissue density has to be included to obtain reliable SAR distributions. For instance, an automated segmentation could be used to assign literature values for tissue density.

If this method should be used for SAR assessment prior to potentially hazardous imaging sequences, the observed underestimation of SAR could be compensated by introducing a suitable safety margin. Using EPT-based SAR determination, a suitable safety margin of at least 50 % in head local SAR measurements at 1.5 T should be taken into account. A routine application could particularly be based on the phase-based approach due to its potentially very short acquisition time. Since the local SAR imaging scan itself will cause SAR, it should be made sure that it is e.g. a low flip angle and / or long TR scan to fulfill SAR limits. Once the SAR pattern is known, the B_p amplitude can be scaled to exploit regulatory limits and hence, the examination could be conducted with optimal RF power. The method is applicable as input for hot spot reduction methods such as [135, 136].

The presented results keep it ambiguous whether the neglection of H_m or H_z is responsible for the observed SAR underestimation. The underestimation seems to depend on the subject's geometry and dielectric properties, and probably the RF coil used. An investigation of the relative importance of each of these factors could be the topic of a future study. For the human brain, H_m seems to be more relevant for accurate SAR determination than H_z (see Section 4.4.1). This feature might turn out as happenstance, since studies to determine H_m are on the way, e.g., in the framework of parallel RF transmission [92]. See appendix C.1 for a more detailed investigation of the impact of H_m and H_z on electric fields and hence SAR in symmetric objects using a plane wave model.

A (maybe approximate) determination of H_z and H_m might become even more important when applying the presented method to field strengths higher than 1.5 T. Here, H_p is no longer the dominant component, and neglecting the other field components could have a severe impact on the quality of the reconstructed SAR. Thus, the development of methods to measure these might become indispensable. A thorough investigation of high-field local SAR measurements using the presented method is postponed to a separate study.

Erroneous SAR distributions can be caused not only by neglected field components, but also by an imperfect measurement of H_p. In the presented study, MTM B_p mapping and eddy current corrected CLEAR spin echo phase φ_p imaging yielded satisfying results and were validated using FDTD simulations. In principle, any sufficiently fast, reliable, and accurate B_p and φ_p mapping method can be used. For instance, Bloch Siegert

(BS) B_p mapping (see [19] or Section 1.4.1) might yield better SNR than MTM mapping; on the other hand, BS mapping has never been tested for 3D B_p mapping capabilities. This is especially critical since it is expected that the BS pulse interacts with the imaging sequence. For phase imaging, FFE sequences could be used in combination with B_0 inhomogeneity correction methods. This approach might be extremely fast, however, also prone to several artifacts occurring in this framework [12].

In general, more beneficial than SAR mapping would be a direct estimation of the RF induced temperature increase [137], since altered temperature is the actual source of potential tissue damage. Besides local SAR, tissue temperature is also determined by cooling mechanisms, heat diffusion, and other parameters [138]. Thus, any temperature model has to include a variety of components, however, will also benefit from an accurate determination of local SAR as suggested in this study.

As all imaging based methods, the presented local SAR measurement technique is only able to detect hot spots inside the FOV. Therefore, the FOV should be set as large as possible to cover the entire coil sensitivity area. In some cases, such as in ankle imaging using the body coil, local SAR hot spots even outside the body coil have been reported [125]. Such situation would have to be excluded from application of the presented method.

4.6 Conclusion

Local SAR measurements using standard MR sequences and suitable post-processing is possible. Considering the suggested safety margins, the presented method yields sufficiently accurate SAR estimates for the relevant regulatory limits such as maximum 10 g local SAR. Calculation times are tremendously shorter than those of FDTD simulations, which can be taken as today's gold standard of local SAR determination. Moreover, it provides a patient-specific local SAR estimate based on measurements that automatically include system imperfections normally not taken into account in simulations. On the other hand, proof of concept in the more challenging but also more interesting field of very high-field and/or parallel transmission systems still has to be delivered.

4.7 Acknowledgments

The authors of [139] would like to thank Astrid van Lier for help on the eddy current correction and Ingmar Graesslin for valuable comments and discussion.

CHAPTER FIVE

SUMMARY, CONCLUSIONS, AND FURTHER DEVELOPMENTS

Si me hallas en error en un solo punto, no debes por eso condenarme en todos.

— MIGUEL SERVET (1511 – 1553)

Advances in MR imaging sequences as well as MR system hardware have led to a broadening of applications since the first medical MR application in the 1970s. The range of possible contrasts as well as the achievable image quality increased substantially. Recently, clinical quantitative imaging gained increasing interest due to its prospect of overcoming the subjectiveness of individual interpretation. Other aspects, beneficial for therapy follow up, are given by the additional dimension offered by the absolute scale. Changes can be monitored based on reliable and reproducible absolute numbers. Quantitative imaging means using the MR scanners as a scientific measurement instrument instead of just "taking pictures", i.e., producing relative, qualitative images.

Various quantitative techniques require the excitation of magnetization to be constant over the relevant FOV. Therefore, additional excitation field correction is frequently performed to obtain correct quantitative values. To this goal, an additional scan is applied to obtain a map of the excitation field, which is then used to correct the base line measurement. In this work, a new method for simultaneous B_p/T_1 mapping called MTM has been introduced. The method yields a relaxometry map, which is automatically corrected for excitation field inhomogeneities. This is achieved by the simultaneous mapping of B_p via the analytic description of the signal of a spoiled steady-state sequence with arbitrary number of different repetition times.

The B_p mapping capabilities of MTM have been compared with AFI, a gold standard method for B_p mapping. The new approach is also able to substantially decrease systematic and statistical B_p mapping errors compared to standard AFI. In this work, MTM has been based on sequences with multiple TR pairs. In future work, also MTM sequences based on TR triples shall be implemented and tested. Simulations showed that TR triples are particular suitable for MTM, i.e., beneficial in terms of SNR per unit time of the reconstructed B_p map.

The T_1 mapping capabilities of MTM were compared with reference measurements and revealed a high degree of accuracy, although speed and SNR still leave room for improvement compared to highly optimized reference sequences. Future work shall be focused on this aspect, e.g., combining MTM with EPI readout and/or undersampling schemes. A prominent clinical application, which could be facilitated by MTM, is DCE-MRI. DCE-MRI is widely applied to assess tissue perfusion and vascular permeability [140]. In most of these applications (see section 1.4.5), a T_1 map, obtained prior to contrast agent (CA) administration, is used to convert the signal intensity from a series of T_1-weighted spoiled gradient-echo images into CA concentration. The accuracy of the approach is mainly affected by the flip angle dependency of the steady-state signal, which necessitates careful quantification of B_p inhomogeneities in advance. Hence, MTM seems to be particularly suitable for this application. Future work shall be aimed at this topic.

Besides existing parameters that have been measured with MR before, two new parameters have been introduced in this thesis. The tissues' dielectric properties, i.e., the conductivity and permittivity, are quantitative parameters accessible in MRI using the presented EPT reconstruction algorithm. EPT is based on the fact that dielectric samples distort the applied spin excitation field. The character of this distortion is determined by the specific, local conductivity and permittivity. The exact relation between dielectric properties and excitation field distortions can be derived using Maxwell's theory of electromagnetic fields. In this work, a thorough validation and first *in vivo* images of healthy volunteers have been obtained with EPT. The new method is able to measure quantitative dielectric properties to a high degree of accuracy. Magnitude-based permittivity imaging and phase-based conductivity imaging have been introduced and successfully tested. Both methods yield significant increase in imaging speed. Phase-based conductivity imaging seems to be mature for clinical evaluation.

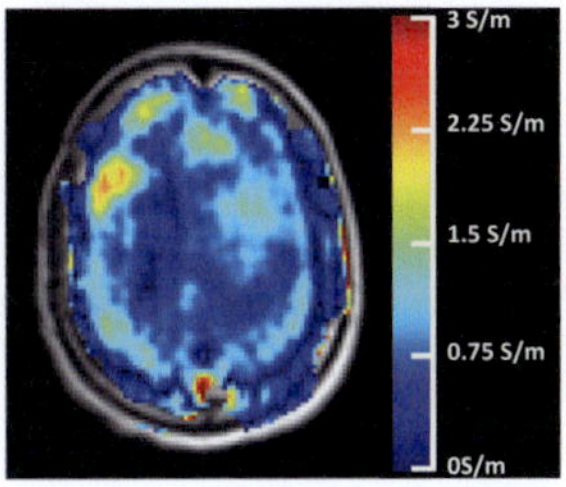

(a)

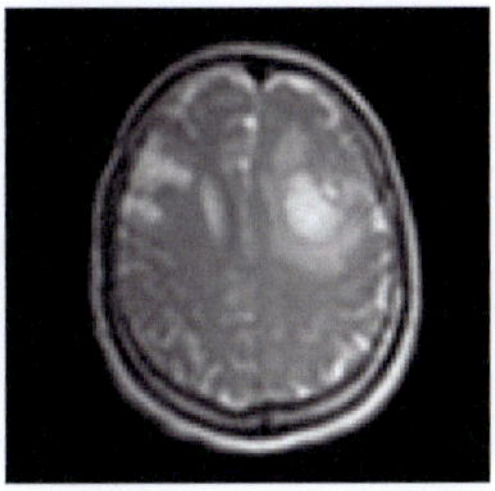

(b)

Figure 5.1: (a) shows a brain conductivity image. A glioblastoma is visible, which shows clearly enhanced conductivity compared to the surrounding white matter. On the opposite side, an already operated region is filled with CSF. (b) shows a corresponding TSE image. φ_p maps courteously provided by Dr. Ole Väterlein (Klinik und Poliklinik für Neuroradiologische Diagnostik und Intervention, Universitätsklinikum Hamburg-Eppendorf).

Thus, a major future task is the clinical evaluation of the presented EPT methods. Interesting from a clinical point of view is the application of EPT in oncology diagnosis, e.g., investigating tumor conductivities and its relation to biochemistry or physiologic parameters like vascularization. Other examples of pathologically altered conductivities might be found in the white matter lesions connected with multiple sclerosis or in the heart muscle suffering atrial fibrillation or infarction. Unfortunately, at the time of finalizing this thesis, systematic clinical studies have just started, which could allow a more precise prediction on possible clinical applications. Fig. 5.1 shows a first conductivity image of a brain tumor. Permittivity quantitative imaging as presented in this work does not seem to be equivalently mature at this point of development. However, applying permittivity imaging at 7 T seems to be promising due to its described beneficial dependence on Larmor Frequency.

The knowledge of tissue's dielectric properties not only supports clinical applications, but also several research applications. For instance, dielectric properties are needed as input to localize electric sources in the area of electro- / magnetoencephalogram (EEG/MEG) or electro- / magnetocardiogram (ECG/MCG) procedures (albeit at frequencies below Larmor). In telecommunications, local SAR computations based on dielectric properties distributions are regularly performed to assess the local heating caused by the use of cell phones.

RF safety in connection with elevated local SAR values is not only a problem in telecommunications, but also in high-field MRI. Currently, clinical high-field systems operate with RF fields capable of causing significant heating of tissue. Therefore, global and local SAR has to be estimated and validated prior to scanning, and sequence parameters have to be adjusted accordingly to ensure patient safety. In this work, a new method for SAR estimation based on post-processing of MR images is introduced. It allows an approximate local SAR reconstruction from measured field distributions, and hence, provides a patient-specific SAR map. It is expected that this map is more reliable than the currently applied local SAR estimations based on standardized patient models, which represent the actual patient variations rather poorly. Continuing the promising studies at 1.5 T presented in this work, future work shall address applications at higher main fields, where local SAR levels turn out to be major problems. Any ideas contributing to cope with these problems will be of central importance for the future of MRI.

List of Figures

List of Tables

Bibliography

The purpose of computing is insight, not numbers.

— RICHARD W. HEMMING (1915 – 1998)

[1] I.I. Rabi, J.R. Zacharias, S. Millman, P. Kusch, "A new method of measuring nuclear magnetic moment", *Physical Review*, vol. 53, no. 4, pp. 318–318, 1938.

[2] F. Bloch, W.W. Hansen, M.D. Packard, "Nuclear introduction", *Physical Review*, vol. 69, pp. 127–132, 1946.

[3] E.M. Purcell, H.C. Torrey, R.V. Pound, "Resonance absorption by nuclear magnetic moments in a solid", *Physical Review*, vol. 69, no. 1-2, pp. 37–38, 1946.

[4] R.V. Damadian, "Tumor detection by nuclear magnetic resonance", *Science*, vol. 171, pp. 1151–1153, 1971.

[5] G.N. Hounsfield, "Computerized transverse axial scanning (tomography): Part 1. Description of system", *British Journal of Radiology*, vol. 46, no. 552, p. 1016, 1973.

[6] P.C. Lauterbur, "Image formation by induced local interactions: examples employing nuclear magnetic resonance", *Nature*, vol. 242, no. 5394, pp. 190–191, 1973.

[7] A. Kumar, D. Welti, R.R. Ernst, "NMR Fourier zeugmatography", *Journal of Magnetic Resonance (1969)*, vol. 18, no. 1, pp. 69–83, 1975.

[8] R. Damadian, L. Minkoff, M. Goldsmith, M. Stanford, J. Koutcher, "Field focusing nuclear magnetic resonance (FONAR): visualization of a tumor in a live animal.", *Science*, vol. 194, no. 4272, p. 1430, 1976.

[9] P. Mansfield, "Multi-planar image formation using NMR spin echoes", *Journal of Physics C: Solid State Physics*, vol. 10, p. L55, 1977.

[10] R. Damadian, M. Goldsmith, L. Minkoff, "NMR in cancer: XVI. FONAR image of the live human body.", *Physiological Chemistry and Physics*, vol. 9, no. 1, pp. 97–100, 1977.

[11] W. Nolting, *Grundkurs Theoretische Physik 5/2: Quantenmechanik-Methoden und Anwendungen*, Springer Verlag, 2006.

[12] EM. Haacke, RW. Brown, MR. Thompson, R. Venkatesan, *Magnetic resonance imaging: physical principles and sequence design*, New York: Wiley-Liss, 1999.

[13] "Guidance for industry and FDA staff. Criteria for significant risk investigations of magnetic resonance diagnostic devices", 2003.

[14] E.L. Hahn, "Spin echoes", *Physical Review*, vol. 80, p. 580, 1950.

[15] H.Y. Carr & E.M. Purcell, "Effects of diffusion on free precession in nuclear magnetic resonance experiments", *Physical Review*, vol. 94, no. 3, pp. 630–638, 1954.

[16] O. Dössel, *Bildgebende Verfahren in der Medizin: von der Technik zur medizinischen Anwendung*, Springer, 1999.

[17] E. Ashton, "Quantitative MR in multi-center clinical trials", *Journal of Magnetic Resonance Imaging*, vol. 31, no. 2, pp. 279–288, 2010.

[18] P. Tofts, *Quantitative MRI of the brain: measuring changes caused by disease*, John Wiley & Sons Inc, 2003.

[19] L.I. Sacolick, F. Wiesinger, I. Hancu, M.W. Vogel, "B1 mapping by Bloch-Siegert shift", *Magnetic Resonance in Medicine*, vol. 63, no. 5, pp. 1315–1322, 2010.

[20] F. Bloch & A. Siegert, "Magnetic resonance for nonrotating fields", *Physical Review*, vol. 57, no. 6, pp. 522–527, 1940.

[21] D.C. Look & D.R. Locker, "Time saving in measurement of NMR and EPR relaxation times", *Review of Scientific Instruments*, vol. 41, pp. 250–251, 1970.

[22] S. Meiboom & D. Gill, "Modified spin-echo method for measuring nuclear relaxation times", *Review of Scientific Instruments*, vol. 29, p. 688, 1958.

[23] D.L. Thomas, M.F. Lythgoe, G.S. Pell, F. Calamante, R.J. Ordidge, "The measurement of diffusion and perfusion in biological systems using magnetic resonance imaging", *Physics in Medicine and Biology*, vol. 45, p. R97, 2000.

[24] R. Venkatesan, W. Lin, E.M. Haacke, "Accurate Determination of Spin-Density and T1 in the Presence of RF-Field Inhomogeneities and Flip-Angle Miscalibration", *Magnetic Resonance in Medicine*, vol. 40, pp. 592–602, 1998.

[25] J.G. Sled & G.B. Pike, "Correction for B1 and B0 variations in quantitative T2 measurements using MRI", *Magnetic Resonance in Medicine*, vol. 43, pp. 589–593, 2000.

[26] J. Wang, M. Qiu, H. Kim, R.T. Constable, "T1 measurements incorporating flip angle calibration and correction in vivo", *Journal of Magnetic Resonance*, vol. 182, no. 2, pp. 283–292, 2006.

[27] R.S.. Samson, C.A.M. Wheeler-Kingshott, M.R. Symms, D.J. Tozer, P.S. Tofts, "A simple correction for B-1 field errors in magnetization transfer ratio measurements", *Magnetic Resonance Imaging*, vol. 24, pp. 255–263, 2006.

[28] T.S. Ibrahim, R. Lee, B.A. Baertlein, A. Kangarlu, P.L. Robitaille, "Application of finite difference time domain method for the design of birdcage RF head coils using multi-port excitations", *Magnetic Resonance Imaging*, vol. 18, pp. 733–42, 2000.

[29] U. Katscher, P. Börnert, C. Leussler, J.S. van den Brink, "Transmit SENSE", *Magnetic Resonance in Medicine*, vol. 49, pp. 144–150, 2003.

[30] E.K. Insko & L. Bolinger, "Mapping of radiology field", *Journal of Magnetic Resonance, Series A*, vol. 103, pp. 82–85, 1993.

[31] R. Stollberger & P. Wach, "Imaging of the Active B1 Field in Vivo", *Magnetic Resonance in Medicine*, vol. 35, pp. 246–251, 1995.

[32] C.H. Cunningham, J.M. Pauly, K.S. Nayak, "Saturated double-angle method for rapid B_1+ mapping", *Magnetic Resonance in Medicine*, vol. 55, pp. 1326–1333, 2006.

[33] J. Wang, M. Qiu, R.T. Constable, "In Vivo Method for Correcting Transmit/Receive Nonuniformities with Phased Array Coils", *Magnetic Resonance in Medicine*, vol. 53, pp. 666–674, 2005.

[34] M.A. Fernández-Seara, H.K. Song, F.W. Wehrli, "Trabecular Bone Volume Fraction Mapping by Low-Resolution MRI", *Magnetic Resonance in Medicine*, vol. 46, pp. 103–113, 2001.

[35] J.P. Hornak, J. Szumowski, R.G. Bryant, "Magnetic field mapping", *Magnetic Resonance in Medicine*, vol. 6, pp. 158–163, 1988.

[36] M. Alecci, C.M. Collins, M.B. Smith, P. Jezzard, "Radio frequency magnetic field mapping of a 3 Tesla birdcage coil: experimental and theoretical dependence on sample properties", *Magnetic Resonance in Medicine*, vol. 46, no. 2, pp. 379–385, 2001.

[37] F. Jiru & U. Klose, "Fast 3D radiofrequency field mapping using echo-planar imaging", *Magnetic Resonance in Medicine*, vol. 56, no. 6, pp. 1375–1379, 2006.

[38] J.W. Carlson & D.M. Kramer, "Fast 3D radiofrequency field mapping using echo-planar imaging", *Magnetic Resonance in Medicine*, vol. 56, pp. 1375–1379, 2006.

[39] S. Akoka, F. Franconi, F. Seguin, A. Le Pape, "Radiofrequency map of an NMR Coil by Imaging", *Magnetic Resonance Imaging*, vol. 11, pp. 437–441, 1993.

[40] N.G. Dowell & P.S. Tofts, "Fast, accurate, and precise mapping of the RF field in vivo using the 180° signal null", *Magnetic Resonance in Medicine*, vol. 58, no. 3, pp. 622–630, 2007.

[41] G. Morrell, "A Phase-Sensitive Method of Flip Angle Mapping", *Magnetic Resonance in Medicine*, vol. 60, pp. 889–894, 2008.

[42] V.L. Yarnykh, "Actual flip-angle imaging in the pulsed steady state: a method for rapid three-dimensional mapping of the transmitted radiofrequency field", *Magnetic Resonance in Medicine*, vol. 57, no. 1, pp. 192–200, 2007.

[43] K. Nehrke, "On the Steady-State Properties of Actual Flip Angle Imaging (AFI)", *Magnetic Resonance in Medicine*, vol. 61, pp. 84–92, 2009.

[44] V.L. Yarnykh, "Optimal spoiling of the transverse magnetization in the Actual Flip-angle Imaging (AFI) sequence for fast B1 field mapping", in *Proceedings of the 16th Annual Meeting of ISMRM*, p. 3090, 2008.

[45] A. Lutti, C. Hutton, J. Finsterbusch, G. Helms, N. Weiskopf, "Optimization and validation of methods for mapping of the radiofrequency transmit field at 3T", *Magnetic Resonance in Medicine*, 2010.

[46] R. Treier, A. Steingoetter, M. Fried, W. Schwizer, P. Boesiger, "Optimized and combined T1 and B1 mapping technique for fast and accurate T1 quantification in contrast-enhanced abdominal MRI", *Magnetic Resonance in Medicine*, vol. 57, no. 3, pp. 568–576, 2007.

[47] J.J. Hsu, G. Zaharchuk, G.H. Glover, "Rapid Methods for Concurrent Measurement of the RF-Pulse Flip Angle and the Longitudinal Relaxation Time", *Magnetic Resonance in Medicine*, vol. 61, pp. 1319–1325, 2009.

[48] T. Voigt, U. Katscher, K. Nehrke, O. Doessel, "Simultaneous B1 and T1 mapping based on modified "Actual Flip-angle Imaging"", in *Proceedings of the 17th Annual Meeting of ISMRM*, p. 4543, 2009.

[49] D. Marquardt, "An Algorithm for least-squares estimation of nonlinear parameters", *Journal of the Society for Industrial and Applied Mathematics*, vol. 11, pp. 431–441, 1963.

[50] SM. Kay, *Fundamentals of Statistical Signal Processing: Estimation Theory*, Englewood Cliffs, NJ: Prentice Hall, 1993.

[51] WH. Press, SA. Teukolsky, WT. Vetterling, BP. Flannery, *Numerical Recipes: the art of scientific computing*, 3 ed., Cambridge University Press, 2007, Ch. Downhill simplex method in multidimensions, pp. 408–412.

[52] J.J. In den Kleef & J.J. Cuppen, "RLSQ: T1, T2 and ρ Calculations, Combining Ratios and Least Squares", *Magnetic Resonance in Medicine*, vol. 5, pp. 513–524, 1987.

[53] P. Thévenaz, U.E. Ruttimann, M. Unser, "A Pyramid Approach to Subpixel Registration Based on Intensity", *IEEE Transactions on Image Processing*, vol. 7, pp. 27–41, 1998.

[54] P.A. Bottomley, T.H. Foster, R.E. Argersinger, L.M. Pfeifer, "A review of normal tissue hydrogen NMR relaxation times and relaxation mechanisms from 1–100 MHz: dependence on tissue type, NMR frequency, temperature, species, excision, and age", *Medical Physics*, vol. 11, p. 425, 1984.

[55] V.L. Yarnykh, "Comparison between RF spoiling schemes in the Actual Flip-angle Imaging (AFI) sequence for fast B1 mapping", in *Proceedings of the 18th Annual Meeting of ISMRM*, p. 241, 2010.

[56] V.L. Yarnykh, "Optimal radiofrequency and gradient spoiling for improved accuracy of T1 and B1 measurements using fast steady-state techniques", *Magnetic Resonance in Medicine*, vol. 63, no. 6, pp. 1610–1626, 2010.

[57] K.P. Pruessmann, M. Weiger, M.B. Scheidegger, P. Boesiger, "SENSE: sensitivity encoding for fast MRI", *Magnetic Resonance in Medicine*, vol. 42, no. 5, pp. 952–962, 1999.

[58] K. Nehrke & P. Börnert, "Improved B1-mapping for multi RF transmit systems", in *Proceedings of the 16th Annual Meeting of ISMRM*, p. 353, 2008.

[59] D.O. Brunner & K.P. Pruessmann, "A matrix approach for mapping array transmit fields in under a minute", in *Proceedings of the 16th Annual Meeting of ISMRM*, p. 354, 2008.

[60] T. Voigt, K. Nehrke, O. Doessel, U. Katscher, "T1 Corrected B1 Mapping Using Multi-TR Gradient Echo Sequences", *Magnetic Resonance in Medicine*, vol. 64, pp. 725–733, 2010.

[61] E.M. Haacke, L.S. Petropoulos, E.W. Nilges, D.H. Wu, "Extraction of conductivity and permittivity using magnetic resonance imaging", *Physics in Medicine and Biology*, vol. 36, pp. 723–734, 1991.

[62] U. Katscher, T. Dorniok, C. Findeklee, P. Vernickel, K. Nehrke, "In vivo determination of electric conductivity and permittivity using a standard MR system", in *IFMBE Proceedings ICEBI*, vol. 17, Springer, p. 508, 2007. 13th International Conference on Electrical Bio-Impedance & 8th Conference on Electrical Impedance Tomography.

[63] U. Katscher, T. Voigt, C. Findeklee, P. Vernickel, K. Nehrke, O. Doessel, "Determination of Electric Conductivity and Local SAR via B1 Mapping", *IEEE Transactions on Medical Imaging*, vol. 28, pp. 1365–1374, 2009.

[64] L.F. Fuks, M. Cheney, D. Isaacson, D.G. Gisser, J.C. Newell, "Detection and imaging of electric conductivity and permittivity atlow frequency", *IEEE Transactions on Biomedical Engineering*, vol. 38, no. 11, pp. 1106–1110, 1991.

[65] B.M. Eyuboglu, T.C. Pilkington, P.D. Wolf, "Estimation of tissue resistivities from multiple-electrode impedance measurements", *Physics in Medicine and Biology*, vol. 39, pp. 1–17, 1994.

[66] P.M. Edic, G.J. Saulnier, J.C. Newell, D. Isaacson, "A real-time electrical impedance tomograph", *IEEE Transactions on Biomedical Engineering*, vol. 42, no. 9, pp. 849–859, 1995.

[67] V. Cherepenin, A. Karpov, A. Korjenevsky, V. Kornienko, A. Mazaletskaya, D. Mazourov, D. Meister, "3D EIT system for breast cancer detection", *Physiological Measurement*, vol. 22, pp. 9–18, 2001.

[68] A.V. Korzhenevskii & V.A. Cherepenin, "Magnetic Induction Tomography", *Journal of Communications Technology & Electronics*, vol. 42, no. 4, pp. 469–474, 1997.

[69] H. Griffiths, "Magnetic induction tomography", *Measurement Science and Technology*, vol. 12, pp. 1126–1131, 2001.

[70] Y. Xu & B. He, "Magnetoacoustic tomography with magnetic induction (MAT-MI)", *Physics in Medicine and Biology*, vol. 50, p. 5175, 2005.

[71] X. Li, Y. Xu, B. He, "Imaging electrical impedance from acoustic measurements by means of magnetoacoustic tomography with magnetic induction (MAT-MI)", *IEEE Transactions on Biomedical Engineering*, vol. 54, no. 2, pp. 323–330, 2007.

[72] S. Oh, A.G. Webb, T. Neuberger, B.S. Park, C.M. Collins, "Experimental and numerical assessment of MRI-induced temperature change and SAR distributions in phantoms and in vivo", *Magnetic Resonance in Medicine*, vol. 63, no. 1, pp. 218–223, 2010.

[73] L.T. Muftuler, M. Hamamura, O. Birgul, O. Nalcioglu, "Resolution and contrast in magnetic resonance electrical impedance tomography (MREIT) and its application to cancer imaging.", *Technology in Cancer Research & Treatment*, vol. 3, no. 6, p. 599, 2004.

[74] E.J. Woo & J.K. Seo, "Magnetic resonance electrical impedance tomography (MREIT) for high-resolution conductivity imaging", *Physiological Measurement*, vol. 29, pp. R1–R26, 2008.

[75] G.C. Scott, M.L.G. Joy, R.L. Armstrong, R.M. Henkelman, "Rotating frame RF current density imaging", *Magnetic Resonance in Medicine*, vol. 33, no. 3, pp. 355–369, 1995.

[76] S. Ueno & N. Iriguchi, "Impedance magnetic resonance imaging: a method for imaging of impedance distributions based on magnetic resonance imaging", *Journal of Applied Physics*, vol. 83, p. 6450, 1998.

[77] M. Sekino, Y. Inoue, S. Ueno, "Magnetic resonance imaging of electrical conductivity in the human brain", *IEEE Transactions on Magnetics*, vol. 41, no. 10, pp. 4203–4205, 2005.

[78] S. Reza, F. Huang, G. Bosman, G.R. Duensing, M.K. Limkeman, C. Saylor, "MR-Noise Tomography: extracting information from noise", in *Proceedings of the 13th Annual Meeting of ISMRM*, p. 955, 2005.

[79] G.R. Duensing, C. Saylor, F. Huang, "Method and apparatus for noise tomography", 2002.

[80] M. Schaefer, W. Gross, J. Ackemann, M.M. Gebhard, "The complex dielectric spectrum of heart tissue during ischemia", *Bioelectrochemistry*, vol. 58, no. 2, pp. 171–180, 2002.

[81] L.X. Liu, W.W. Dong, X.M. Ji, L.H. Chen, L. Chen, W. He, J.P. Jia, "A new method of noninvasive brain-edema monitoring in stroke: cerebral electrical impedance measurement", *Neurological Research*, vol. 28, no. 1, pp. 31–37, 2006.

[82] W.T. Joines, Y. Zhang, C. Li, R.L. Jirtle, "The measured electrical properties of normal and malignant human tissues from 50 to 900 MHz", *Medical Physics*, vol. 21, p. 547, 1994.

[83] A.J. Surowiec, S.S. Stuchly, J.R. Barr, A. Swarup, "Dielectric properties of breast carcinoma and the surrounding tissues", *IEEE Transactions on Biomedical Engineering*, vol. 35, no. 4, pp. 257–263, 1988.

[84] I. Graesslin, K. Falaggis, P. Vernickel, P. Röschmann, C. Leussler, Z. Zhai, M. Morich, U. Katscher, "Safety considerations concerning SAR during RF amplifier malfunctions in parallel transmission", in *Proceedings of the 14th Annual Meeting of ISMRM*, p. 2041, 2006.

[85] A.C. Zelinski, L.M. Angelone, V.K. Goyal, G. Bonmassar, E. Adalsteinsson, L.L. Wald, "Specific Absorption Rate Studies of the Parallel Transmission of Inner-Volume Excitations at 7 Tesla", *Journal of Magnetic Resonance Imaging*, vol. 28, no. 4, p. 1005, 2008.

[86] F. Seifert, G. Wübbeler, S. Junge, B. Ittermann, H. Rinneberg, "Patient safety concept for multichannel transmit coils", *Journal of Magnetic Resonance Imaging*, vol. 26, no. 5, pp. 1315–1321, 2007.

[87] S. Buchenau, M. Haas, J. Hennig, M. Zaitsev, "A comparison of local SAR using individual patient data and a patient template", in *Proceedings of the 17th Annual Meeting of ISMRM*, p. 4798, 2009.

[88] M.A. Cloos & G. Bonmassar, "Towards Direct B1 Based Local SAR Estimation", in *Proceedings of the 17th Annual Meeting of ISMRM*, p. 3037, 2009.

[89] S.B. Bulumulla, T.B. Yeo, Y. Zhu, "Direct calculation of tissue electrical parameters from B1 maps", in *Proceedings of the 17th Annual Meeting of ISMRM*, p. 3043, 2009.

[90] U. Katscher, M. Hanft, P. Vernickel, C. Findeklee, "Electric Properties Tomography (EPT) via MRI", in *Proceedings of the 14th Annual Meeting of ISMRM*, Seattle, Washington, USA, p. 3037, 2006.

[91] U. Katscher, M. Hanft, P. Vernickel, C. Findeklee, "Experimental verification of electric properties tomography (ept)", in *Proceedings of the 14th Annual Meeting of ISMRM*, 2006.

[92] U. Katscher, C. Findeklee, T. Voigt, "Experimental estimation of local SAR in a multi-transmit system", in *Proceedings of the 17th Annual Meeting of ISMRM*, p. 4512, 2009.

[93] X. Zhang, S. Zhu, B. He, "Imaging Electric Properties of Biological Tissues by RF Field Mapping in MRI", *IEEE Transactions on Medical Imaging*, vol. 2, pp. 474–481, 2010.

[94] X. Zhang & B. He, "Imaging electric properties of human brain tissues by B1 mapping: A simulation study", in *Journal of Physics: Conference Series*, vol. 224, IOP Publishing, p. 012077, 2010.

[95] H. Wen, "Noninvasive quantitative mapping of conductivity and dielectric distributions using RF wave propagation effects in high-field MRI", in *Proceedings of SPIE*, vol. 5030, p. 471, 2003.

[96] T. Voigt, O. Doessel, U. Katscher, "Imaging conductivity and local SAR of the human brain", in *Proceedings of the 17th Annual Meeting of ISMRM*, p. 4513, 2009.

[97] S. Gabriel, R.W. Lau, C. Gabriel, "The dielectric properties of biological tissues: III. Parametric models for the dielectric spectrum of tissues", *Physics in Medicine and Biology*, vol. 41, pp. 2271–2293, 1996.

[98] H.P. Schwan & K. Li, "Capacity and conductivity of body tissues at ultrahigh frequencies", *Proceedings of the IRE*, vol. 41, no. 12, pp. 1735–1740, 1953.

[99] K.R. Foster & H.P. Schwan, "Dielectric properties of tissues and biological materials: A critical review", *Critical Reviews in Biomedical Engineering*, vol. 17, pp. 25–104, 1989.

[100] H.P. Schwan, "Electrical properties of tissue and cell suspensions.", *Advances in biological and medical physics*, vol. 5, pp. 147–209, 1957.

[101] B.R. Epstein & K.R. Foster, "Anisotropy in the dielectric properties of skeletal muscle", *Medical and Biological Engineering and Computing*, vol. 21, no. 1, pp. 51–55, 1983.

[102] J.D. Jackson, *Classical electrodynamics*, John Wiley. NY US, 1999.

[103] D.I. Hoult, "The principle of reciprocity in signal strength calculations - a mathematical guide", *Concepts in Magnetic Resonance*, vol. 12, no. 4, pp. 173–187, 2000.

[104] A.L. Van Lier, A.J. Raaijmakers, D.O. Brunner, D.W. Klomp, K.P. Pruessmann, J.J. Lagendijk, C.A. van den Berg, "Propagating RF Phase: A New Contrast to Detect Local Changes in Conductivity", in *Proceedings of the 18th Annual Meeting of ISMRM*, Stockholm, Sweden, p. 2864, 2010.

[105] MNO. Sadiku, *Numerical Techniques in Electro-magnetics*, CRC Press LLC, 2001.

[106] A. Savitzky & M.J.E. Golay, "Smoothing and differentiation of data by simplified least squares procedures.", *Analytical chemistry*, vol. 36, no. 8, pp. 1627–1639, 1964.

[107] C. Gabriel & A. Peyman, "Dielectric measurement: error analysis and assessment of uncertainty", *Physics in Medicine and Biology*, vol. 51, p. 6033, 2006.

[108] T. Voigt, U. Katscher, O. Doessel, "In vivo Quantitative Conductivity Imaging based on B1 Phase Information", in *Proceedings of the 18th Annual Meeting of ISMRM*, Stockholm, Sweden, p. 2865, 2010.

[109] "Particular requirements for the basic safety and essential performance of magnetic resonance equipment for medical diagnosis", 03 2010.

[110] I. Graesslin, D. Gläsel, S. Biederer, P. Vernickel, U. Katscher, F. Schweser, B. Annighoefer, H. Dingemanns, G. Mens, G. v. Yperen, P. Harvey, "Comprehensive RF Safety Concept for Parallel Transmission Systems", in *Proceedings of the 16th Annual Meeting of ISMRM*, 2008.

[111] C.M. Collins & M.B. Smith, "Calculations of B_1 Distribution, SNR, and SAR for a Surface Coil Adjacent to an Anatomically-Accurate Human Body Model", *Magnetic Resonance in Medicine*, vol. 45, no. 4, pp. 692–699, 2001.

[112] I. Graesslin, K. Falaggis, S. Biederer, D. Glaesel, P. Vernickel, P. Roeschmann, C. Leussler, A. Thran, Z. Zhai, M. Morich, U. Katscher, "SAR simulations and experiments for parallel transmission", in *Proceedings of the 15th Annual Meeting of ISMRM*, Berlin, p. 1090, 2007.

[113] K.S. Yee, "Numerical Solution of Initial Boundary Value Problems Involving Maxwell's Equations in Isotropic Media", *IEEE Trans. Antennas Propagat*, vol. 14, no. 3, pp. 302–307, 1966.

[114] C.M. Collins, S. Li, M.B. Smith, "SAR and B_1 Field Distributions in a Heterogeneous Human Head Model within a Birdcage Coil", *Magnetic Resonance in Medicine*, vol. 40, pp. 847–856, 1998.

[115] J. Chen, Z. Feng, J.M. Jin, "Numerical simulation of SAR and B_1-field inhomogeneity of shielded RF coils loaded with the human head", *IEEE Transactions on Biomedical Engineering*, vol. 45, no. 5, pp. 650–659, 1998.

[116] T.S. Ibrahim, R. Lee, B.A. Baertlein, P.M.L. Robitaille, "B1 field homogeneity and SAR calculations", *Physics in Medicine and Biology*, vol. 46, pp. 609–619, 2001.

[117] C.M. Collins & M.B. Smith, "Spatial resolution of numerical models of man and calculated specific absorption rate using the FDTD method: a study at 64 MHz in a magnetic resonance imaging coil", *Journal of Magnetic Resonance Imaging*, vol. 18, no. 3, pp. 383–388, 2003.

[118] W. Liu, C.M. Collins, M.B. Smith, "Calculations of B_1 distribution, specific energy absorption rate, and intrinsic signal-to-noise ratio for a body-size birdcage coil loaded with different human subjects at 64 and 128 MHz", *Applied Magnetic Resonance*, vol. 29, no. 1, pp. 5–18, 2005.

[119] J.W. Hand, Y. Li, J.V. Hajnal, "Numerical study of RF exposure and the resulting temperature rise in the foetus during a magnetic resonance procedure", *Physics in Medicine and Biology*, vol. 55, p. 913, 2010.

[120] J.W. Hand, Y. Li, E.L. Thomas, M.A. Rutherford, J.V. Hajnal, "Prediction of specific absorption rate in mother and fetus associated with MRI examinations during pregnancy", *Magnetic Resonance in Medicine*, vol. 55, no. 4, pp. 883–893, 2006.

[121] Z. Wang, J.C. Lin, W. Mao, W. Liu, M.B. Smith, C.M. Collins, "SAR and temperature: simulations and comparison to regulatory limits for MRI", *Journal of Magnetic Resonance Imaging*, vol. 26, no. 2, pp. 437–441, 2007.

[122] U.D. Nguyen, J.S. Brown, I.A. Chang, J. Krycia, M.S. Mirotznik, "Numerical evaluation of heating of the human head due to magnetic resonance imaging", *IEEE Transactions on Biomedical Engineering*, vol. 51, no. 8, pp. 1301–1309, 2004.

[123] C.A.T. Van den Berg, L.W. Bartels, B. Bergen, H. Kroeze, A.A.C. Leeuw, J.B. Kamer, J.J.W. Lagendijk, "The use of MR B_1^+ imaging for validation of FDTD electromagnetic simulations of human anatomies", *Physics in Medicine and Biology*, vol. 51, pp. 4735–4745, 2006.

[124] National Institutes of Health, *Visible Human Project*, http://www.nlm.nih.gov, 1996. Bethesda, MD, USA.

[125] X. Chen, Y. Hamamura, M. Steckner, "Numerical simulation of SAR for 3T whole body coil: effect of patient loading positions on local SAR hotspot", in *Proceedings of the 18th Annual Meeting of ISMRM*, 2010.

[126] W.D. Hurt, J.M. Ziriax, P.A. Mason, "Variability in EMF Permittivity Values: Implications for SAR Calculations", *IEEE Transactions on Biomedical Engineering*, vol. 47, no. 3, pp. 396–401, 2000.

[127] B. Quesson, J.A. de Zwart, C.T.W. Moonen, "Magnetic resonance temperature imaging for guidance of thermotherapy", *Journal of Magnetic Resonance Imaging*, vol. 12, no. 4, pp. 525–533, 2000.

[128] E.M. Shapiro, A. Borthakur, M.J. Shapiro, R. Reddy, J.S. Leigh, "Fast MRI of RF heating via phase difference mapping", *Magnetic Resonance in Medicine*, vol. 47, no. 3, pp. 492–498, 2002.

[129] P.A. Bottomley, R.W. Redington, W.A. Edelstein, J.F. Schenck, "Estimating radiofrequency power deposition in body NMR imaging", *Magnetic Resonance in Medicine*, vol. 2, no. 4, pp. 336–349, 1985.

[130] Matt A. Bernstein, Kevin F. King, Xiaohong J. Zhou, *Handbook of MRI Pulse Sequences*, 1 ed., Academic Press, September 2004.

[131] "IEEE recommended practice for measurements and computations of radio frequency electromagnetic fields with respect to human exposure to such fields, 100 kHz-300 GHz", 2002.

[132] S.B. Reeder, Z. Wen, H. Yu, A.R. Pineda, G.E. Gold, M. Markl, N.J. Pelc, "Multicoil Dixon chemical species separation with an iterative least-squares estimation method", *Magnetic Resonance in Medicine*, vol. 51, no. 1, pp. 35–45, 2004.

[133] H. Homann, I. Graesslin, T. Voigt, P. Börnert, O. Dössel, "Towards patient-specific SAR models: A study on spatial and dielectric model resolution", in *MAGMA 22*, p. 144, 2009.

[134] C. Gabriel, S. Gabriel, E. Corthout, "The dielectric properties of biological tissues: I. Literature survey", *Physics in Medicine and Biology*, vol. 41, pp. 2231–2249, 1996.

[135] C.A.T. Van den Berg, B. van den Bergen, J.B. Van de Kamer, B.W. Raaymakers, H. Kroeze, L.W. Bartels, J.J.W. Lagendijk, "Simultaneous B_1^+ Homogenization and Specific Absorption Rate Hotspot Suppression Using a Magnetic Resonance Phased Array Transmit Coil", *Magnetic Resonance in Medicine*, vol. 57, pp. 577–586, 2007.

[136] I. Graesslin, C. Steiding, B. Annighoefer, J. Weller, D. Brunner, H. Homann, F. Schweser, U. Katscher, K. Pruessmann, P. Börnert, "Local SAR constrained Hotspot Reduction by Temporal Averaging", in *Proceedings of the 18th Annual Meeting of ISMRM*, p. 4932, 2010.

[137] C.M. Collins, W. Liu, J. Wang, R. Gruetter, J.T. Vaughan, K. Ugurbil, M.B. Smith, "Temperature and SAR calculations for a human head within volume and surface coils at 64 and 300 MHz", *Journal of Magnetic Resonance Imaging*, vol. 19, no. 5, pp. 650–656, 2004.

[138] H.H. Pennes, "Analysis of tissue and arterial blood temperatures in the resting human forearm", *Journal of Applied Physiology*, vol. 1, no. 2, p. 93, 1948.

[139] T. Voigt, H. Homann, U. Katscher, O. Doessel, "Patient-specific in vivo local SAR Estimation and Validation", in *Proceedings of the 18th Annual Meeting of ISMRM*, Stockholm, Sweden, p. 3876, 2010.

[140] A.R. Padhani, "Dynamic contrast-enhanced MRI in clinical oncology: current status and future directions", *Journal of Magnetic Resonance Imaging*, vol. 16, no. 4, pp. 407–422, 2002.

Appendix

A MTM B_p/T_1 mapping

A.1 Error propagation theory for $N = 3$ generalized AFI

For the performance comparison of standard AFI and generalized AFI using $N = 3$ presented in chapter 2 an analytical expression for error propagation in Eq. 2.6 is required. In this appendix Gaussian error propagation will be applied to Eq. 2.6, which can be written as

$$\alpha_{a,b} = \arccos\left(-\frac{p}{2} \pm \sqrt{\frac{p^2}{4} - q}\right) \tag{A.1}$$

where the following definitions have been made

$$p = \frac{r_{i,j}\mathrm{TR}_{\mathrm{mod}(j,N)+1} - \mathrm{TR}_{\mathrm{mod}(i,N)+1}}{\left(r_{i,j}\mathrm{TR}_{\mathrm{mod}(j-1,N)+1} - \mathrm{TR}_{\mathrm{mod}(i-1,N)+1}\right)}, \tag{A.2}$$

$$q = \frac{r_{i,j}\mathrm{TR}_{\mathrm{mod}(j+1,N)+1} - \mathrm{TR}_{\mathrm{mod}(i+1,N)+1}}{r_{i,j}\mathrm{TR}_{\mathrm{mod}(j-1,N)+1} - \mathrm{TR}_{\mathrm{mod}(i-1,N)+1}}, \text{ and} \tag{A.3}$$

$$B = -\frac{p}{2} \pm \sqrt{\frac{p^2}{4} - q}. \tag{A.4}$$

Error propagation theory is then applied to the function $\alpha_{i,j}$ (Eq. A.1) of the two measured signals S_i and S_j. The variance ς^2 then reads

$$\varsigma^2 = \varsigma_{\mathrm{in}}^2 \left(\frac{\partial \alpha_{i,j}}{\partial S_i} + \frac{\partial \alpha_{i,j}}{\partial S_j}\right)^2 \tag{A.5}$$

where ς_{in} denotes the standard deviation of the input signals $S_{i,j}$. Now the derivatives with respect to the signals have to be computed:

$$\frac{\partial \alpha_{i,j}}{\partial S_i} = \frac{-1}{\sqrt{1-B^2}}\frac{\partial B}{\partial S_i} \text{ and } \frac{\partial \alpha_{i,j}}{\partial S_j} = \frac{-1}{\sqrt{1-B^2}}\frac{\partial B}{\partial S_j} \tag{A.6}$$

with

$$\begin{aligned} \frac{\partial B}{\partial S_i} &= \frac{-\partial p}{2\partial S_i} \pm \frac{1}{2}\left(\frac{p^2}{4}-q\right)^{-\frac{1}{2}}\left[\frac{p}{2}\frac{\partial p}{\partial S_i}-\frac{\partial q}{\partial S_i}\right] \\ \frac{\partial B}{\partial S_j} &= \frac{-\partial p}{2\partial S_j} \pm \frac{1}{2}\left(\frac{p^2}{4}-q\right)^{-\frac{1}{2}}\left[\frac{p}{2}\frac{\partial p}{\partial S_j}-\frac{\partial q}{\partial S_j}\right] \end{aligned} \tag{A.7}$$

where

$$\begin{aligned} \frac{\partial p}{\partial S_i} = & \frac{\frac{1}{S_j}\mathrm{TR}_{\mathrm{mod}(j,N)+1}\left(r_{i,j}\mathrm{TR}_{\mathrm{mod}(j-1,N)+1}-\mathrm{TR}_{\mathrm{mod}(i-1,N)+1}\right)}{\left(r_{i,j}\mathrm{TR}_{\mathrm{mod}(j-1,N)+1}-\mathrm{TR}_{\mathrm{mod}(i-1,N)+1}\right)^2} \\ & -\frac{\left(r_{i,j}\mathrm{TR}_{\mathrm{mod}(j,N)+1}-\mathrm{TR}_{\mathrm{mod}(i,N)+1}\right)\frac{1}{S_j}\mathrm{TR}_{\mathrm{mod}(j-1,N)+1}}{\left(r_{i,j}\mathrm{TR}_{\mathrm{mod}(j-1,N)+1}-\mathrm{TR}_{\mathrm{mod}(i-1,N)+1}\right)^2} \end{aligned} \tag{A.8}$$

$$\begin{aligned} \frac{\partial p}{\partial S_j} = & \frac{\frac{-S_i}{S_j^2}\mathrm{TR}_{\mathrm{mod}(j,N)+1}\left(r_{i,j}\mathrm{TR}_{\mathrm{mod}(j-1,N)+1}-\mathrm{TR}_{\mathrm{mod}(i-1,N)+1}\right)}{\left(r_{i,j}\mathrm{TR}_{\mathrm{mod}(j-1,N)+1}-\mathrm{TR}_{\mathrm{mod}(i-1,N)+1}\right)^2} \\ & +\frac{\left(r_{i,j}\mathrm{TR}_{\mathrm{mod}(j,N)+1}-\mathrm{TR}_{\mathrm{mod}(i,N)+1}\right)\frac{S_i}{S_j^2}\mathrm{TR}_{\mathrm{mod}(j-1,N)+1}}{\left(r_{i,j}\mathrm{TR}_{\mathrm{mod}(j-1,N)+1}-\mathrm{TR}_{\mathrm{mod}(i-1,N)+1}\right)^2} \end{aligned} \tag{A.9}$$

$$\begin{aligned} \frac{\partial q}{\partial S_i} = & \frac{\frac{1}{S_j}\mathrm{TR}_{\mathrm{mod}(j+1,N)+1}\left(r_{i,j}\mathrm{TR}_{\mathrm{mod}(j-1,N)+1}-\mathrm{TR}_{\mathrm{mod}(i-1,N)+1}\right)}{\left(r_{i,j}\mathrm{TR}_{\mathrm{mod}(j-1,N)+1}-\mathrm{TR}_{\mathrm{mod}(i-1,N)+1}\right)^2} \\ & -\frac{\frac{1}{S_j}\mathrm{TR}_{\mathrm{mod}(j-1,N)+1}\left(r_{i,j}\mathrm{TR}_{\mathrm{mod}(j+1,N)+1}-\mathrm{TR}_{\mathrm{mod}(i+1,N)+1}\right)}{\left(r_{i,j}\mathrm{TR}_{\mathrm{mod}(j-1,N)+1}-\mathrm{TR}_{\mathrm{mod}(i-1,N)+1}\right)^2} \end{aligned} \tag{A.10}$$

$$\frac{\partial q}{\partial S_j} = \frac{\frac{-S_i}{S_j^2}\mathrm{TR}_{\mathrm{mod}(j+1,N)+1}\left(r_{i,j}\mathrm{TR}_{\mathrm{mod}(j-1,N)+1} - \mathrm{TR}_{\mathrm{mod}(i-1,N)+1}\right)}{\left(r_{i,j}\mathrm{TR}_{\mathrm{mod}(j-1,N)+1} - \mathrm{TR}_{\mathrm{mod}(i-1,N)+1}\right)^2} + \frac{\frac{S_i}{S_j^2}\mathrm{TR}_{\mathrm{mod}(j-1,N)+1}\left(r_{i,j}\mathrm{TR}_{\mathrm{mod}(j+1,N)+1} - \mathrm{TR}_{\mathrm{mod}(i+1,N)+1}\right)}{\left(r_{i,j}\mathrm{TR}_{\mathrm{mod}(j-1,N)+1} - \mathrm{TR}_{\mathrm{mod}(i-1,N)+1}\right)^2} \tag{A.11}$$

The expressions Eqs. A.6 to A.11 can now be inserted into the formula for error propagation, Eq. A.5.

A.2 Phantom experiment with long T_1 and T_2

In the phantom study presented in section 2.3.3.1 it has been shown that with maximum TR of about 1/3 of maximum T_1, the systematic T_1 dependent error is negligible. To exclude MTM as reason for the occurring CSF artifacts (Fig. 2.11a), additional phantom experiments were performed. The phantom consisted of two compartments with $T_1 \approx 3\text{s}/0.2\text{s}$. MTM B_p and T_1 maps were obtained using the same imaging parameters as in the *in vivo* case described in the manuscript. Results are shown in Tab. A.1. It can be deduced that MTM is able to fully compensate systematic AFI errors even for very long T_1 and T_2, as expected from theory. Therefore, it is assumed that the elevation visible in CSF B_p in Fig. 2.11a is due to other effects, e.g. CSF flow.

B Electric properties tomography

B.1 Derivation of EPT formalism using differential representation

The derivations leading to the EPT equations displayed in integral representation in section 3.3.1 can also be executed in differential representation. From a mathematical point of view, there is no difference between the two approaches and favoring one over the other is a mere question of preference. In differential representation Ampere's law reads

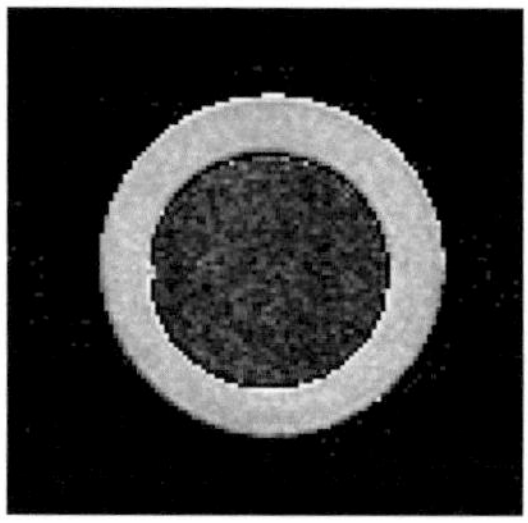

Figure A.1: Difference in reconstructed flip angle between MTM and AFI $\Delta\alpha = \alpha_{\mathrm{MTM}} - \alpha_{\mathrm{AFI}}$ [°]. The difference is clearly T_1 dependent. A quantitative analysis is given in Tab. A.1.

	T_1		$\Delta\alpha = \alpha_{\mathrm{MTM}} - \alpha_{\mathrm{AFI}}$		T_2
	From RLSQ	From MTM	measured	simulation	RLSQ
Inner	3010 ± 57 ms	3018 ± 161 ms	$0.09 \pm 0.13°$	$0.062°$	1576 ± 188 ms
Outer	231 ± 48 ms	238 ± 9 ms	$1.67 \pm 0.23°$	$1.64°$	200 ± 15 ms

Table A.1: Results from two-compartment phantom experiment with long T_1 and T_2. The differences between MTM and AFI are in agreement with simulation results, indicating that MTM not only reduces, but fully compensates T_1 errors even for TR $\ll T_1$.

$$(i\omega)^{-1}\nabla\times\mathbf{H}(\mathbf{r})=\kappa(\mathbf{r})\mathbf{E}(\mathbf{r}) \tag{B.1}$$

where $\mathbf{H},\mathbf{E}\sim\exp(i\omega t)$ has been used. Applying $\nabla\times$ to both sides, using the vector identity $\nabla\times\nabla\times\mathbf{H}=\nabla(\nabla\cdot\mathbf{H})-\triangle\mathbf{H}=-\triangle\mathbf{H}$, and assuming constant dielectric properties κ yields

$$-(i\omega)^{-1}\triangle\mathbf{H}=\kappa(\mathbf{r})\nabla\times\mathbf{E}(\mathbf{r}) \tag{B.2}$$

Dividing this equation by Faraday's law leads to the differential version of Eq. 3.4

$$\frac{-\triangle\mathbf{H}}{\omega^2\mu\mathbf{H}}=\kappa(\mathbf{r})\,. \tag{B.3}$$

Applying the same coordinate transformation as in section 3.3.1 yields the differential version of Eq. 3.10

$$\frac{-\triangle H_p}{\omega^2\mu H_p}=\kappa(\mathbf{r})\,. \tag{B.4}$$

Also the approximated, separated conductivity and permittivity Eqs. 3.19 and 3.20 can be expressed in differential versions

$$\sigma(\mathbf{r}) \approx (\omega\mu_0 V)^{-1}\triangle\varphi_p, \tag{B.5}$$

$$\varepsilon(\mathbf{r}) \approx \frac{-\triangle B_p(\mathbf{r})\mathrm{d}\mathbf{a}}{\omega^2\mu_0 B_p(\mathbf{r})}. \tag{B.6}$$

B.2 Orthogonality of coordinate system

The orthogonality of the used circularly polarized coordinate system is proven in this part of the appendix. Proof for the vectors $\mathbf{e}_p=\frac{1}{\sqrt{2}}(\mathbf{e}_x+i\mathbf{e}_y)$ and $\mathbf{e}_m=\frac{1}{\sqrt{2}}(\mathbf{e}_x-i\mathbf{e}_y)$:

$$\begin{aligned}
\langle\mathbf{e}_p,\mathbf{e}_m\rangle &= \tfrac{1}{\sqrt{2}}\langle(\mathbf{e}_x+i\mathbf{e}_y),\mathbf{e}_m\rangle=\tfrac{1}{\sqrt{2}}\left[\langle\mathbf{e}_x,\mathbf{e}_m\rangle+\langle i\mathbf{e}_y,\mathbf{e}_m\rangle\right]\\
&= \tfrac{1}{2}\left[\langle\mathbf{e}_x,(\mathbf{e}_x-i\mathbf{e}_y)\rangle+\langle i\mathbf{e}_y,(\mathbf{e}_x-i\mathbf{e}_y)\rangle\right]\\
&= \tfrac{1}{2}\left[\langle\mathbf{e}_x,\mathbf{e}_x\rangle+\langle\mathbf{e}_x,-i\mathbf{e}_y\rangle+\langle i\mathbf{e}_y,\mathbf{e}_x\rangle+\langle i\mathbf{e}_y,-i\mathbf{e}_y\rangle\right]\\
&= \tfrac{1}{2}\left[\langle\mathbf{e}_x,\mathbf{e}_x\rangle-i\langle\mathbf{e}_x,\mathbf{e}_y\rangle-i\langle\mathbf{e}_y,\mathbf{e}_x\rangle+(-i)^2\langle\mathbf{e}_y,\mathbf{e}_y\rangle\right]\\
&= \tfrac{1}{2}\left[\langle\mathbf{e}_x,\mathbf{e}_x\rangle-\langle\mathbf{e}_y,\mathbf{e}_y\rangle\right]=0.
\end{aligned} \tag{B.7}$$

The corresponding proofs for $\langle \mathbf{e}_p, \mathbf{e}_z \rangle$ and $\langle \mathbf{e}_m, \mathbf{e}_z \rangle$ are trivial and not shown explicitly.

C SAR imaging

C.1 Investigating the significance of electric field components in symmetric objects

Differently oriented plane waves are used to discuss the significance of certain electric and magnetic field components in symmetric objects. First, a plane wave travelling in z-direction is regarded, which can be described by $\mathbf{H} = \mu^{-1}\mathbf{B}e^{ikz}$ with $B = (B_p, B_m, B_z)^T$. Its derivative with respect to x and y components are zero $\partial_p H_i = 0$ and $\partial_m H_i = 0$. Hence, the electric field reads (cf. Eq. 4.10)

$$\begin{aligned} i\omega\kappa(\mathbf{r})E_p(\mathbf{r}) &\approx i\partial_z H_p(\mathbf{r}), \\ i\omega\kappa(\mathbf{r})E_m(\mathbf{r}) &\approx -i\partial_z H_m(\mathbf{r}), \\ i\omega\kappa(\mathbf{r})E_z(\mathbf{r}) &\approx 0. \end{aligned} \tag{C.1}$$

H_z does not enter the electric fields in this model. Hence, if wave propagation in patients is predominantly in z-direction, the z-component of the magnetic field H_z does not influence the electric fields to a large extent, and thus, will not have significant influence on local SAR as observed in the FDTD studies of Chapter 4.

On the other hand, for the upright cylinders used in the phantom experiment, the symmetry axis is the anterior-posterior direction (y-direction). For a plane wave travelling in y-direction, the relations $\partial_x H_i = 0$ and $\partial_z H_i = 0$ hold. Inserting these relations into Eq. 4.10, the electric fields become

$$\begin{aligned} i\omega\kappa(\mathbf{r})E_p(\mathbf{r}) &\approx \partial_y H_z(\mathbf{r}), \\ i\omega\kappa(\mathbf{r})E_m(\mathbf{r}) &\approx \partial_y H_z(\mathbf{r}), \\ i\omega\kappa(\mathbf{r})E_z(\mathbf{r}) &\approx -\partial_y H_m(\mathbf{r}) - \partial_y H_p(\mathbf{r}). \end{aligned} \tag{C.2}$$

Thus, H_z plays a more important role in this scenario. H_m enters the electric fields only in one component, leading to a smaller influence on electric fields (and thus, local SAR) than H_z. This is in accordance with observations from phantom experiments of Chapter 4.

D Translation of quotations

In this part of the appendix, an attempt of translating the quotes preceding each chapter into English language is made.

Nil actum reputans si quid superesset agendum. - Carl Friedrich Gauß	Nothing has been done if something remains to be done.
Scienza è il distinguere quello che si sa da quello che non si sa. - Galileo Galilei	Science is separating what you know from what you do not know.
Wär nicht das Auge sonnenhaft, die Sonne könnt es nie erblicken. - Johann Wolfgang von Goethe	If the eye were not sun-like, the sun's light it would not see.
Dans la vie, rien n´est à craindre, tout est à comprendre. - Marie Skłodowska-Curie	Nothing in life is to be feared, it is only to be understood.
Insofern sich die Sätze der Mathematik auf die Wirklichkeit beziehen, sind sie nicht sicher, und insofern sie sicher sind, beziehen sie sich nicht auf die Wirklichkeit. - Albert Einstein	As far as the laws of mathematics refer to reality, they are not certain; and as far as they are certain, they do not refer to reality.

Publications

The secret is comprised in three words - work, finish, publish.

— MICHAEL FARADAY (1791 – 1867)

Publications in International Journals (full papers):

- Voigt T, Homann H, Doessel O, Katscher U. Patient-Individual Local SAR Determination: *In Vivo* Measurements and FDTD Validation. Under review in Magn Reson Med.
- Voigt T, Katscher U, Doessel O. Quantitative Conductivity and Permittivity Imaging of the Human Brain using Electric Properties Tomography. Magn Reson Med 2010, in press.
- Voigt T, Nehrke K, Doessel O, Katscher U. T1 corrected B1 mapping Using Multi-TR Gradient Echo Sequences. Magn Reson Med 2010;64:725-733.
- Katscher U, Voigt T, Findeklee C, Vernickel P, Nehrke K, Dössel O. Determination of Electric Conductivity and Local SAR via B1 Mapping. IEEE Trans Med Imaging 2009;28:1365-1374.

Publications in International Conference Proceedings:

- Voigt T, Homann H, Katscher U, Doessel O. Patient-specific in vivo local SAR Estimation and Validation. In Proceedings of the 18th Annual Meeting of ISMRM, Stockholm, Sweden 2010. p. 3876.
- Voigt T, Katscher U, Doessel O. In vivo Quantitative Conductivity Imaging based on B1 Phase Information. In Proceedings of the 18th Annual Meeting of ISMRM, Stockholm, Sweden 2010. p. 2865.
- Voigt T, Remmele S, Katscher U, Doessel O. Fast T1/B1 Mapping using multiple dual TR RF-spoiled Gradient Echo Sequences. In Proceedings of the 18th Annual Meeting of ISMRM, Stockholm, Sweden 2010. p. 2949.
- Katscher U, Voigt T, Findeklee C. Estimation of the Anisotropy of Electric Conductivity via B1 Mapping. In Proceedings of the 18th Annual Meeting of ISMRM, Stockholm, Sweden 2010. p. 2866.

- Katscher U, Karkowski P, Findeklee C, Voigt T. Permittivity determination via phantom and in vivo B1 mapping. In Proceedings of the 18th Annual Meeting of ISMRM, Stockholm, Sweden 2010. p. 239.

- Remmele S, Voigt T, Keupp J, Stehning C, Sénégas J. Simultaneous ΔR1 and ΔR2* Quantification in 5s to Monitor Blood and Tissue Oxygenation with Dynamic (C)O2 Enhanced MRI. In Proceedings of the 18th Annual Meeting of ISMRM, Stockholm, Sweden 2010. p 5121.

- Katscher U, Voigt T, Findeklee C. Electrical conductivity imaging using magnetic resonance tomography. In Proceedings of 31st Annual International Conference of the IEEE EMBS (2009), Minneapolis, USA. p. 3162- 3164.

- Homann H, Graesslin I, Voigt T, Börnert P, Dössel O. Towards patient-specific SAR models: A study on spatial and dielectric model resolution. MAGMA 22 (2009). p. 144.

- Homann H, Graesslin I, Voigt T, Börnert P, Dössel O. The influence of body size on the specific absorption rate (SAR). MAGMA 22 (2009). p. 386.

- Voigt T, Katscher U, Findeklee C, Dössel O. Imaging Electric Conductivity with MRI. In IFMBE Proceedings World Congress on Medical Physics and Biomedical Engineering 25/II (2009). pp 42.

- Voigt T, Doessel O, Katscher U. Imaging conductivity and local SAR of the human brain. In Proceedings of the 17th Annual Meeting of ISMRM, Honolulu, Hawaii, USA 2009. p. 4513.

- Voigt T, Katscher U, Nehrke K, Doessel O. Simultaneous B1 and T1 mapping based on modified "Actual Flip-angle Imaging". In Proceedings of the 17th Annual Meeting of ISMRM, Honolulu, Hawaii, USA 2009. p. 4543.

- Katscher U, Findeklee C, Voigt T. Experimental estimation of local SAR in a multi-transmit system. In Proceedings of the 17th Annual Meeting of ISMRM, Honolulu, Hawaii, USA 2009. p. 4512.

- Voigt T, Nehrke K, Dössel O, Katscher U. In vivo determination of local SAR. MAGMA 21 (2008) p 65-66.

- Voigt T, Findeklee C, Dössel O, Katscher U. Imaging electric properties with MR. MAGMA 21 (2008). p400-401.

- Katscher U, Voigt T, Findeklee C, Nehrke K, Weiss S, Doessel O. Estimation of local SAR using B1 mapping. In Proceedings of the 16th Annual Meeting of ISMRM, Toronto, Canada 2008. p. 1191.

Patents:

- Remmele S, Voigt T, Stehning C, Liu W. Simultaneous and dynamic determination of longitudinal and transversal relaxation behavior of a nuclear spin system. US Patent 61/315062.
- Voigt T, Katscher U, Homann H. A method and a program for calculating local specific energy absorption rate (SAR) and a magnetic resonance system. European Patent 09181034.1.
- Katscher U, Voigt T, Harvey P, Homann H, Findeklee C, Hansis E. Electric properties tomography imaging Method and system. European Patent 10150979.2.
- Nehrke K, Boernert P, Voigt T, Katscher U. Flip angle imaging with improved B1 mapping for multi-rf transmit systems, WO Patent WO/2009 / 118,702.
- Weiss, S, Katscher U, Vernickel P, Voigt T, Findeklee C. Determination of local SAR in vivo and electrical conductivity mapping, WO Patent WO/2009 / 118,688.

Zusammenfassung

Insofern sich die Sätze der Mathematik auf die Wirklichkeit beziehen, sind sie nicht sicher, und insofern sie sicher sind, beziehen sie sich nicht auf die Wirklichkeit.

— ALBERT EINSTEIN (1879 – 1955)

In der letzten Zeit haben quantitative Bildgebungsverfahren in der Magnetresonanztomographie (MRT) an Bedeutung gewonnen. Im Unterschied zur klassischen, anatomischen Bildgebung ist mit quantitativen Verfahren die Hoffnung verbunden, die klinische Aussage von MRT Untersuchungen objektiver gestalten zu können. Quantitative Messgrößen können einer bestimmten Region zugeordnet werden und ermöglichen so die Messung einer bestimmten Eigenschaft des Gewebes. Krankhafte Gewebeveränderungen oder Veränderungen in Folge von Therapien werden zu einer Änderung dieser Messgröße führen. Eine Aussage kann somit unabhängig von der individuellen Interpretation eines Radiologen getroffen werden und sollte ebenfalls von anderen äußeren Einflüssen, wie z.B. nicht idealen Messbedingungen, unabhängig sein.

Die vorliegende Arbeit widmet sich der Entwicklung neuer quantitativer Bildgebungsmethoden. Nach einer Einleitung in die MR Physik sowie einem Überblick über eine Reihe in der Forschung etablierter, quantitativer Messverfahren beschäftigt sich das zweite Kapitel mit der Messung des Protonenanregungsfeldes. Eine neue Methode zur simultanen Messung der Amplitude des Anregungsfeldes B_p und der longitudinalen Relaxationszeit T_1 wird vorgestellt. Die neue Methode, genannt "Multiple TR B_p/T_1 Mapping" (MTM), basiert auf einer Erweiterung der bereits existierenden "Actual Flip-angle Imaging" (AFI)-Methode. Die erweiterte Sequenz beinhaltet anstelle von zwei Repetitionszeiten (TR) ein beliebige Anzahl TRs. In der vorliegenden Arbeit wird eine analytische Beschreibung der sich ergebenden Magnetisierung gefunden. Die MTM-Methode wird außerdem mit der herkömmlichen AFI-Messung verglichen. MTM liefert dabei B_p Karten mit höherem Signal-Rausch-Verhältnis. Eine Machbarkeitsstudie zur T_1 Bestimmung mit MTM wird ebenfalls erfolgreich durchgeführt; hier zeigt sich, dass MTM in der Lage ist, Referenzmessungen mit hoher Genauigkeit zu reproduzieren.

Im dritten Kapitel wird eine quantitative Messmethode zur Bestimmung der dielektrischen Eigenschaften von Gewebe, "Electric Properties Tomography" (EPT), beschrieben und weiterentwickelt. Die beiden dielektrischen Eigenschaften von Gewebe, d.h. seine Leitfähigkeit und Permittivität bewirken eine Verzerrung des (im Falle einer leeren Spule konstanten) Anregungsfeldes. Mittels der Maxwellschen Theorie für elektromagnetische Wellen lässt sich eine analytische Rekonstruktionsformel für die dielektrischen Eigenschaften als Funktion einer einzelnen Komponente des Anregungsfeldes herleiten. Basierend auf einer Messung dieser einen Komponente des Anregungsfeldes wird die resultierende neue EPT Methode in dieser Arbeit mittels Phantomstudien validiert und Leitfähigkeits- und Permittivitätsmessungen *in vivo* am menschlichen Gehirn durchgeführt. Darauf aufbauend beinhaltet diese Arbeit außerdem eine Weiterentwicklung von EPT, die separate, phasenbasierte Leitfähigkeitsmessungen und amplitudenbasierte Permittivitätsmessungen ermöglicht und somit eine teilweise erhebliche Steigerung der Geschwindigkeit der Messungen erlaubt. Diese Methoden werden theoretisch motiviert, durch Simulationen validiert und in *in vivo* Versuchen getestet. Bei diesen Untersuchungen stellt sich das weiterentwickelte EPT als außerordentlich erfolgreich und leistungsfähig heraus.

Im vierten Kapitel werden Feldmessungen und Messungen der dielektrischen Eigenschaften zur Berechnung der lokalen spezifischen Absorptionsrate (SAR) benutzt. Regulatorische Gremien der EU und der USA bestimmen Grenzwerte für die lokale SAR, die während einer MRT Untersuchung nicht überschritten werden dürfen. Bei intensiver Hochfrequenz-Einstrahlung während einer MRT Untersuchung kann es zur unerwünschten Erwärmung und eventuell Verbrennung von Gewebe kommen. Das lokale SAR muss daher vor jeder MRT Sequenz abgeschätzt und die Sequenzparameter eventuell angepasst werden. In dieser Arbeit wird eine Methode zur lokalen SAR Messung vorgestellt und anhand von Simulationen, Phantommessungen und schließlich im *in vivo* Experiment getestet. Es wird gezeigt, dass die neue Methode eine hinreichend genaue, patientenspezifische Abschätzung der lokalen SAR liefern kann. Es handelt sich weltweit um die erste Studie, die eine derartige *in vivo* Bestimmung der lokalen SAR mit Hilfe von MRT untersucht.

Karlsruhe Transactions on Biomedical Engineering (ISSN 1864-5933)

Karlsruhe Institute of Technology / Institute of Biomedical Engineering (Ed.)

Die Bände sind unter www.ksp.kit.edu als PDF frei verfügbar oder als Druckausgabe bestellbar.

Band 2 Matthias Reumann
Computer assisted optimisation on non-pharmacological treatment of congestive heart failure and supraventricular arrhythmia. 2007
ISBN 978-3-86644-122-4

Band 3 Antoun Khawaja
Automatic ECG analysis using principal component analysis and wavelet transformation. 2007
ISBN 978-3-86644-132-3

Band 4 Dmytro Farina
Forward and inverse problems of electrocardiography : clinical investigations. 2008
ISBN 978-3-86644-219-1

Band 5 Jörn Thiele
Optische und mechanische Messungen von elektrophysiologischen Vorgängen im Myokardgewebe. 2008
ISBN 978-3-86644-240-5

Band 6 Raz Miri
Computer assisted optimization of cardiac resynchronization therapy. 2009
ISBN 978-3-86644-360-0

Band 7 Frank Kreuder
2D-3D-Registrierung mit Parameterentkopplung für die Patientenlagerung in der Strahlentherapie. 2009
ISBN 978-3-86644-376-1

Band 8 Daniel Unholtz
Optische Oberflächensignalmessung mit Mikrolinsen-Detektoren für die Kleintierbildgebung. 2009
ISBN 978-3-86644-423-2

Band 9 Yuan Jiang
Solving the inverse problem of electrocardiography in a realistic environment. 2010
ISBN 978-3-86644-486-7

Karlsruhe Transactions on Biomedical Engineering (ISSN 1864-5933)

Band 10 Sebastian Seitz
Magnetic Resonance Imaging on Patients with Implanted Cardiac Pacemakers. 2011
ISBN 978-3-86644-610-6

Band 11 Tobias Voigt
Quantitative MR Imaging of the Electric Properties and Local SAR based on Improved RF Transmit Field Mapping. 2011
ISBN 978-3-86644-598-7